INTERVIEWING
THE BASICS

This text outlines the relative merits of qualitative interviewing to new and emerging scholars in an accessible way. This is achieved not by providing an exhaustive 'how-to' guide but by introducing researchers to the interview technique and using examples of 'best practice' from across the social sciences.

To ensure the book is both accessible and inclusive, efforts have been made to include case studies from a diverse range of authors, including those from different ethnic and social backgrounds, from outside Western Europe/North America, and from non-academic sources. This book will therefore introduce the reader to the key themes surrounding interview design, implementation, analysis and presentation, using examples and case studies from research across the social sciences.

Crucially, the book will not provide exhaustive guidance on how to conduct the techniques. Instead, each chapter includes a range of interview design activities for readers to try which might help them engage with the chapter topics, as well as a 'Summary' box which comprises a short annotated reading list of key texts relating to each of the chapter topics and a checklist of things to consider relating to interview design, practice and presentation.

Mark Holton is a human geographer who focuses on the social and cultural geographies of young people and youth culture. Mark has considerable experience in designing and utilising conventional and experimental qualitative techniques across a range of externally funded research projects. This has resulted in a variety of methods-related outputs for leading academic journals (e.g., *Area* and *Mobilities*); chapters for academic reference titles (e.g., *Handbook of Qualitative Research in Education* and *The SAGE Encyclopaedia of Higher Education*); and online research methods guides (e.g., *Oxford Bibliographies* and *SAGE Research Methods*). Most recently, Mark co-edited *Creative Methods for Human Geographers*, a 29-chapter book that introduces, mainly novice, researchers to a range of creative approaches to research design.

THE BASICS

The Basics is a highly successful series of accessible guidebooks which provide an overview of the fundamental principles of a subject area in a jargon-free and undaunting format.

Intended for students approaching a subject for the first time, the books both introduce the essentials of a subject and provide an ideal springboard for further study. With over 50 titles spanning subjects from artificial intelligence (AI) to women's studies, The Basics are an ideal starting point for students seeking to understand a subject area. Each text comes with recommendations for further study and gradually introduces the complexities and nuances within a subject.

For a full list of titles in this series, please visit www.routledge.com/The-Basics/book-series/B

INTERVIEWING
THE BASICS

Mark Holton

Routledge
Taylor & Francis Group

LONDON AND NEW YORK

Designed cover image: ©Getty image

First published 2025
by Routledge
4 Park Square, Milton Park, Abingdon, Oxon OX14 4RN

and by Routledge
605 Third Avenue, New York, NY 10158

Routledge is an imprint of the Taylor & Francis Group, an informa business

© 2025 Mark Holton

The right of **Mark Holton** to be identified as author of this work has been asserted in accordance with sections 77 and 78 of the Copyright, Designs and Patents Act 1988.

British Library Cataloguing-in-Publication Data
A catalogue record for this book is available from the British Library

Library of Congress Cataloging-in-Publication Data
Names: Holton, Mark (Lecturer in human geography), author.
Title: Interviewing : the basics / Mark Holton.
Description: Abingdon, Oxon; New York, NY: Routledge, 2025. |
Series: The basics | Includes bibliographical references and index. |
Identifiers: LCCN 2024014308 (print) | LCCN 2024014309 (ebook) |
ISBN 9781032274393 (hardback) | ISBN 9781032274409 (paperback) |
ISBN 9781003292784 (ebook)
Subjects: LCSH: Interviewing in sociology. | Human geography. | Interviewing.
Classification: LCC HM571 .H66 2025 (print) |
LCC HM571 (ebook) | DDC 001.4/33--dc23/eng/20240614
LC record available at https://lccn.loc.gov/2024014308
LC ebook record available at https://lccn.loc.gov/2024014309

ISBN: 978-1-032-27439-3 (hbk)
ISBN: 978-1-032-27440-9 (pbk)
ISBN: 978-1-003-29278-4 (ebk)

DOI: 10.4324/9781003292784

Typeset in Bembo
by SPi Technologies India Pvt Ltd (Straive)

CONTENTS

LISTS OF ILLUSTRATIONS

ACKNOWLEDGEMENTS

I would like to thank Andrew Mould from Routledge for approaching me to write this book. This has been a volume that I have theoretically 'written' for the last decade or so through working with interview-based research methods in my own research, as well as with hundreds of undergraduate and postgraduate students. I therefore would like to dedicate this to the colleagues who have taught me, the students who have worked with me and the research participants who have let me listen to their wonderful stories.

PREFACE – ABOUT QUALITATIVE INTERVIEWING

STATEMENT OF AIMS

Titles on specific research methods are rare in the Routledge Basics library, so this book aims to showcase the relative merits of qualitative interviewing to new and emerging scholars in a detailed and accessible way. I caveat this by stating that this will be achieved not by providing an exhaustive 'how-to' guide (although the 'how' is generously attended to in the following chapters), but instead, I will introduce researchers to the interviewing technique and use examples of 'best practice' from research conducted across the breadth of the social sciences.

This book's primary aim is to introduce new researchers to interviewing techniques, and through the following chapters, I will encourage readers to

- **understand** the basic principles of what interviewing is and why it is a useful and appropriate method;
- **strategise** how best to plan and design an interview-led research project;
- **engage** with interview practice using hints, tips and advice from a range of social science research; and
- **experiment** with new and emerging techniques within the field to help researchers tailor their methods to their own projects.

WHO IS THIS BOOK FOR?

When preparing this book, I considered that it might be likely for it to have a varied audience. I certainly do not want to privilege one research context over another, and, as listed next, I write with a diverse audience in mind. Hence, I have structured and written this book in a way that should appeal to both academic and non-academic readers. I have tried, where possible, to use accessible language, define technical terminology, explain techniques and practices, and showcase examples of best practices. The book's audience might include the following:

- Sixth-form/high school students who may be conducting research-led projects (e.g., the Extended Project Qualification (EPQ) in the United Kingdom) that require primary research to be carried out
- University students embarking on methods training for the first time who may be preparing to conduct primary research as part of fieldwork modules or for their dissertations
- Postgraduate students unfamiliar with qualitative techniques
- Natural scientists who may be interested in drawing qualitative approaches into their research design
- Non-academic practitioners or researchers who may be interested in utilising interviews in their practice but outside of an educational setting

While this book is primarily geared towards supporting readers from, or interested in, the social sciences, there may be subsidiary interest for other scholars from the natural sciences who have interests in adopting cross-disciplinary methods into their research design. While this is not compulsory in many educational settings, cross-disciplinary research is becoming increasingly popular among researchers tackling important global problems. Moreover, when reading this book, it might not surprise you to learn that my academic discipline is Human Geography. I certainly approached writing this book, rather unapologetically I might add, using my knowledge of space, place, scale and time to help understand and contextualise some of the environmental and interactive dimensions of interview design, practice and analysis, yet, in doing so, I have made explicit reference to a range of exciting, innovative and accessible research from social scientists and non-academic practitioners who attend to space and place when interviewing.

HOW TO READ THIS BOOK

I would, of course, presume that the majority of readers picking up this book are doing so in order to gain knowledge for academic research projects. Yet, while this book will speak easily towards an academic research audience, its content is also relatable to a variety of non-academic and professional contexts that seek to elicit knowledge, opinions or feedback from participant groups. The chapters are structured so as to be read independently or chronologically, meaning readers can navigate the text depending on their needs and/or experience, and I include a glossary to help define key terminology.

Providing an overview of interviewing could inevitably produce quite a hefty tome. So, in designing this book, it was therefore crucial that I found the right balance between piquing a reader's curiosity in considering and/or adopting interview techniques into their research and providing enough background knowledge and support to help guide them through the process. This book should therefore be treated as an introduction to the key themes surrounding interview design and practice, alongside interview data analysis and presentation, using examples and case studies from research across the social sciences.

Crucially, the book does not provide exhaustive guidance on how to conduct the techniques. Instead, each chapter includes supportive case studies; activities that readers can try which may help them engage with the chapter topics; checklists of things to consider when preparing, conducting, analysing and presenting interview materials; and annotated reading lists of key texts relating to each chapter.

I hope you enjoy the book!

PART I
PREPARING INTERVIEWS

"A CONVERSATION WITH A PURPOSE"?

Chapter objectives

This chapter will introduce you to the qualitative interview as a research method. By reading the chapter, you should

- understand what constitutes an interview and the origins of interview approaches,
- appreciate the importance of interview training in ensuring credibility in interview-based research, and
- recognise the types of knowledge that interviews produce and how these are distinct from other research methods.

INTRODUCTION

"The interview is a conversation with a purpose" (Bingham and Moore, 1931: 3). I open this book with a short quote from the American industrial psychologists Walter Bingham and Bruce Moore. When we interview people, we are attempting to find out information about, and form knowledge of, a particular phenomenon, process or activity that we perhaps have limited, or no, prior understanding of. The way we achieve this is by talking to people – usually those with knowledge, perceptions or experiences of the thing we are investigating. This often sounds rather informal, and there have been significant debates about the 'scientific*ness*' of interviewing practices (see Denzin and Lincoln, 2011; Edwards and Holland, 2013). Interviews are, however, not just casual chats. They have structure and form that allow the interviewer to pose questions

DOI: 10.4324/9781003292784-2

that the interview participant provides responses to. Hence, interviews are, as Bingham and Moore (1931) state, "a conversation with a purpose", in that, if conducted correctly, should be designed to be flexible and fluid whilst also operating within a framework that aims to actively respond to a research problem.

Interviews are fundamentally concerned with understanding human perceptions and experiences. As qualitative researchers, we often choose interviews to gather depth and context to a phenomenon rather than attempting to generalise the broader characteristics of an issue or problem, as with quantitative research (Rubin and Rubin, 2011). Yet, as Donalek (2005) argues, interviews – or, more specifically, the practice(s) of interviewing – can be somewhat paradoxical. Compared to the expert knowledge and skill required to competently and safely operate laboratory equipment or run complex and sophisticated statistical techniques, interviewing a participant sounds deceptively simple. We can all hold a conversation, right? So, should we not all be able to just have a chat with someone about the problem we are trying to solve? Put simply, no. Interviewing, as a research technique, can involve many overlapping processes. For example, an interviewer will have a conversation with a participant, but they will also need to ensure all the themes of the research topic are covered whilst encouraging the participant to talk freely and naturally. They will have to check that the recording equipment works throughout the encounter and make sure the interview participant is comfortable and relaxed. They might make notes during the interview process. They will also need to be responsive to external influences, such as noises or other people that might distract the interview participant. All the while, interviewers will need to appear professional and in control of the situation so the interview participant can settle into that "conversation with a purpose". This might involve handling difficult or probing questions in a calm and sensitive manner, encouraging participants to elaborate on complex issues that require questions being asked in different ways or dealing with unexpected (or even unwanted) responses.

Some may approach this book with preconceived ideas that interviews are unscientific and inferior to quantitative approaches that privilege generalisable and replicable results. This chapter is therefore designed to counter this image and deliver credibility for interviewing as a legitimate and vitally important (and ultimately pretty exciting) research method. The chapter is divided into three sections. First, I will provide a brief overview of what constitutes a qualitative interview, alongside a brief history of the origins of interviewing – from the objectivity/subjectivity debates of the mid-20th century to the cultural turn in the 1980s. Next, I pose the question, "Why interview?" and reflect on the value of understanding the social world to be necessarily complex and disordered, as well as the ways in which qualitative researchers during the cultural turn began to develop 'interview toolkits' to help ensure rigour and validity. Finally, I turn attention to the types of data that interviews produce to help understand how we can represent, and do justice to, the stories that our participants tell us.

WHAT IS AN INTERVIEW?

As I mentioned earlier, interviewing is primarily concerned with understanding human perceptions and experiences. Interviews are always qualitative in design and are subjective by nature. We can, therefore, safely presume the conversational qualities of interview practice as being a form of storytelling in which the questions posed by an interviewer are designed to encourage a fluid and rich narrative from an interview participant (Edwards and Holland, 2013). I use the term 'storytelling' here not to be whimsical or flippant but to intentionally allude to the free-form, individual and unstandardised qualities of interviews that many other quantitative research methods strive to avoid (to get an idea of the richness of interviewing, see Reyes McGovern's (2019) experiences of interviewing Ms. Luna Martinez, a Mexican American teacher about motherhood and education). Where survey questionnaires are designed to follow a rigid, linear and replicable format that requires participants to provide (mainly) specific and standardised responses, interviews require an interactive conversation between the interviewer and interview participant that, whilst scaffolded with questions, consciously encourages participants to provide their own responses in the ways that feel most natural to them (Valentine, 2005). It is important to mention at this point that the types of interview discussions I provide guidance on in this book are one-to-one conversations. This is the most typical form of interview practice and is likely to be the primary approach that you will take for your own research. That said, there are a variety of other interview techniques that involve different groupings of people. You might come across literature that discusses 'paired', 'joint' or 'family' interviews, and these are useful if you want to find out about shared practices, explore relationships and contrasts of people's experiences of an issue together or corroborate, or triangulate, information (see Bjørnholt and Farstad, 2014; Riley, 2014; Sakellariou et al., 2013).

As a practice, interviews have a long lineage as a research method, yet it may come as no surprise that, over the years, interviews have fallen in and out of favour within academic circles – particularly within the social science community (Cochrane, 2020). The interview, as a stand-alone method, is a fairly recent phenomenon, but its roots stem from 1920s sociological and anthropological studies (Edwards and Holland, 2013). In its earlier forms, interviewing was typically subsumed into broader methodological approaches as a way of contextualising research, being used alongside observational techniques, oral histories or oral surveys. Yet, the pioneers of psychoanalysis in the first part of the 20th century saw interviewing as a vital method for delving into the psyche of the human mind (Denzin and Lincoln, 2011). Crucially, interviews broke through the previously objective barriers of 'researcher' and 'subject', facilitating a shift from simply observing behaviour to asking the question, 'Why?' This put interviewing in a precarious place. Whilst it was used as an academic method, the popularity of psychoanalysis saw the commercial potential of interview practices in other spheres, notably market research.

Notwithstanding the popularity of interviewing as a practice, such commerciality did nothing though to bolster the academic credibility of interviewing, and by the mid-point of the 20th century, interviews had fallen out of favour as a methodological approach, being considered, at best, anecdotal, and at worst, unscientific (Cochrane, 2020). At this point, academic research was firmly in the grip of what was known as the 'quantitative revolution' (Denzin and Lincoln, 2011). This period was signified by the quest for truth through objective, replicable and rigorous research approaches – that is, research that can be conducted repeatedly to achieve the same results, thus demonstrating a high degree of reliability (Alston, 1996). Here, science was considered a superior tool for producing knowledge in ways that researchers using discourse (or conversation) were not able to compete with.

Some might argue, then, that the steer towards quantitative, scientific research methods paints a rather gloomy picture of the prospect of interviewing as practice. Whilst this was true for the mid-part of the 20th century, a new cultural revolution began emerging in academic communities during the 1980s and 1990s that was focussed on methods and approaches that could explain, and respond to, the changing social, political and cultural landscapes of capitalist economies and the growing anger at how culture and identity were marginalised in Marxist theories of society (Savage, 2010; Bryman, 1984). Coined the 'qualitative', or 'cultural', turn, this period saw social scientists become increasingly frustrated with quantitative methodologies that were unable to engage with, and understand, individuals' everyday actions, behaviours and routines in sufficient depth (Thrift, 2000). A new raft of research methods, based upon interpretivist ways of understanding the world, began to emerge. Interpretivist approaches are valuable within interview research as they help question the ways in which individuals or cultures create, or interpret, their own version(s) of 'reality' (Ryan, 2018). Hence, at the start of the cultural turn, the data produced through interviews was considered a vital way of interpreting the differences that exist between individuals, groups or cultures in different social, spatial and temporal contexts (Ryan, 2018).

Importantly, these interpretivist ways of thinking about the production of knowledge put the interactions between researchers and participants at the heart of research design and analysis (see Chapter 4 for discussion of positionality and power in research) whilst aligning such stances with feminist, queer, postcolonial, postmodern and post-structural theories to form methodological toolkits that argued for subjectivity, marginalised voices and reflexivity in research (Shurmer-Smith, 2002; Thrift, 2000). In relation to interview practice specifically, this entailed both recognising and explaining the role of the interviewer in the interview process, as well as providing opportunities for groups typically situated at the margins of society to tell their own stories about everyday life. Researchers began to use qualitative techniques, such as interviews, focus groups, ethnography and participant observation, as stand-alone methods rather than as ancillary or corroborative procedures to quantitative research. These methods were used to reach marginalised groups, such as women, sexual and ethnic minority groups, disabled people, children and the elderly, among others. Such qualitative ways of understanding the world switched the focus of research away from objectively researching 'on' people to subjectively working 'with' them to produce

knowledge of their perceptions of, and experiences and engagements with, the world (Rose, 1997; Ray and Sayer, 1999).

WHY INTERVIEW?

As the cultural turn gained momentum, interviewing was soon
viewed as one of the primary qualitative approaches for researchers to engage with. Interviewing provided opportunities for social
scientists to produce knowledge that was in-depth – in that researchers were able to shift beyond investigating how a phenomenon
exists to understand *why* it exists, *who* it affects and *where* differences might apply (McDowell, 1992). In direct contrast with the
scientific community's focus on studies that make generalised claims
about a population and can be replicated in other contexts and by
other researchers, qualitative interviews produce knowledge that is
unashamedly partial, situated, small-scale, derived from small sample
sizes and messy[1] in nature (Law, 2004). The interview, was, therefore, viewed as one of the most appropriate ways of exploring the
complexities of the social world, specifically people's everyday social
lives, because interviewers understood that people are all inherently
different. Acknowledging such differences, therefore, allows the
richness of human behaviour to be explored. Many social scientists
justified these new approaches as removing the façade of 'order' so
as to reveal social life as complex, chaotic and necessarily *dis*ordered.
Indeed, to attempt to order society perhaps risks over-simplifying
the complexity, and range, of human behaviours and rendering marginalised voices as mere outliers.

Yet, as I will elaborate on further throughout this book, it is not
enough for researchers to just head out into the field and start talking
to people. Even during the cultural turn, interviewing still carried
the stigma of the past, and social researchers have had to work hard at
developing a kind of 'methodological scaffolding' around interview
practice to ensure its credibility as a scientific method (James and
Busher, 2006; Flick, 2002). I will talk about ways of determining
rigour and responsibility in interview research in Chapter 3 of this
book, and for qualitative researchers to make interviews a credible,
replicable and sustainable method, they needed to develop interview
design criteria that put interview training and expertise at the heart
of the interviewer's toolkit. Books like this one emerged to inspire
and train budding researchers with the art and skills of qualitative

interviewing (see Bryman, 2016; Silverman, 2015; Robson, 2002). These volumes often encouraged novice researchers to consider the suitability of qualitative interviewing to help answer particular types of research questions and understand how interviews might elicit and produce certain types of knowledge that other methods cannot. Interview-based research projects have since been extolled for their ability to showcase everyday research scenarios, voices and problems that quantitative methods often struggle to make sense of, as well as advocating the subjective power of interviewing to raise awareness of, and for, marginalised voices (see Philo's (1992) work on interviewing rural children in the United Kingdom, and Nagar-Ron and Motzafi-Haller's (2011) interviews with first-generation Mizrahi Jewish women immigrants in Israel). This has crucially involved foregrounding concepts around positionality and power (i.e., who we are as researchers and participants, and the power our respective identities hold in interview encounters) as *part of* research rather than something to be controlled against (see Chapter 4). Moreover, as Wheeler and Morgan Brett (2021) suggest, unlike other methods, interviews are highly responsive in that they can adapt 'in the moment', respond to or encourage unexpected deviations and be followed up after the event. This makes interviewing one of the most intensive and reflexive methods available to researchers, ostensibly because they respond particularly well to the 'how' and 'why' research questions we want to ask our participants.

WHAT DATA DO I WANT TO PRODUCE?

This chapter has, so far, explored the lineage of interviewing and the current centrality of interviewing in the qualitative researcher's toolkit. In this section, I will examine the different types of data that interviewing can produce. At the beginning of this chapter, I outlined that, as a qualitative approach, interviews capture knowledge of people's experiences, perceptions, feelings, emotions and memories. Interviews therefore produce data derived from verbal narratives or conversations. Whilst these very often take the form of *verbal* information[2] that is later transcribed into the *written* word, it is the verbal data – the *spoken words* – that should take precedence in interviews. Unlike other qualitative methods that analyse existing written or visual texts (e.g., discourse analysis and semiotics) or observe behaviours (e.g., ethnography and participant observation), interviews produce verbal data that requires close engagement with what people say and how they say it. It is, therefore, no surprise then that Rubin and Rubin (2011) sub-titled their book on qualitative interviewing *The Art of Hearing Data*. Being a good interviewer requires good listening skills – both listening 'in the moment' to understand, interpret and respond to the conversation in hand (see Chapter 6) and during the transcribing, analysis and writing up phases of the research (see Chapters 9 and 10). When attempting to make sense of participants' words, audio recordings can be extremely valuable in putting you 'back inside' the interview setting to hear precisely what a participant has said and how they have said it, helping to represent these experiences faithfully, carefully and responsibly on the page.

Yet, there is sometimes a tendency for many texts on qualitative interviewing to focus *only* on the spoken word, often moving swiftly from techniques of recording interviews to transcribing and analysing them (see Chapter 9 for advice on transcription and analysis). As Seidman (2006) argues, we, as interviewers, must also attend carefully to a range of other registers in an interview in order to provide rounded context to the data produced. I mentioned earlier that interviews require close engagement with what people say and how they say it, and it can be said that the 'how' often gets overlooked. In Chapter 4, I discuss notions of power and positionality in interviewing contexts to argue that our backgrounds and identities can often play significant roles in how questions are asked, responses are provided and knowledge is received and interpreted. By exploring the other non-verbal, visual and tonal cues that also comprise an interview encounter (see Chapter 7) – a participant's body language; the pitch, tone and tempo of their responses; and intonations, such as passivity, boredom, sarcasm or hostility – we are able to add layers of context to our interviews that ensure they are rich and meaningful data through which to analyse and present our interview participants' perceptions and experiences (Bamberg, 2020).

Moreover, new technologies, such as video conferencing software, social media and mobile phone apps have steered classic in-person interview practice online (see Chapter 8). Researching during the COVID-19 pandemic, for example, presented enormous challenges to interview practice that had, up to 2020, mainly involved in-person human interaction. Social distancing and lockdowns meant that previously standard face-to-face interview practices could not be conducted, meaning researchers had to adapt quickly in order to develop methodologies that could replicate the interactive circumstances of an interview but in an online capacity. Video conferencing software, such as Zoom, Microsoft Teams and Skype, have provided platforms to activate the conversation with a purpose and reproduce, in some ways, many of the interpersonal aspects of traditional interviewing. What has become exciting, though, is the push by many researchers to think about how these platforms might be utilised creatively by using the various functions of these technologies – sharing screens, online whiteboards, games and puzzles, etc. – to capture the more-than-verbal dimensions of an interview. For example, Kobakhidze et al.'s (2021) research on the use of technology in research during the COVID-19 pandemic revealed their Hong Kong–based participants to be extremely adaptive and creative in drawing technology into their existing interview practices. Whilst online practices have endured past the pandemic – largely becoming part of the fundamental 'basics' of contemporary interview practice – when you are designing your own interviews, it is still important to weigh up the potential value of face-to-face interviewing against the perceived convenience of digital practice to ensure you are collecting data that respond directly to your chosen research questions.

SUMMARY POINTS

- Interviews are designed to understand human perceptions, experiences and knowledge.
- Unlike quantitative data – which is objective, standardised and generalisable – qualitative interview data is subjective, partial and often messy.
- Interview data seeks to 'interpret' a phenomenon rather than attempt to find 'the truth'.
- Interview techniques are designed to explore *why* a phenomenon or problem exists.
- Interview approaches are useful in developing knowledge of everyday scenarios, voices and problems and have been used advantageously to raise awareness of, and for, marginalised voices.
- Interviews often privilege the spoken word, but it is also important to recognise the other non-verbal, visual and tonal cues that also comprise an interview encounter to understand precisely what a participant has said and how they have said it.

WHAT TO DO NEXT?

Consider the following:

- First of all, weigh up carefully what types of data you are hoping to produce from an interview and why interviewing is the most appropriate method for you to produce these data.
- Whilst it might not be necessary to immerse yourself in the philosophical foundations of qualitative research, try to remind yourself that interview data is subjective, messy and often partial and consider how this relates to your research ideas.

Suggested further reading

King, N., Horrocks, C., and Brooks, J. (2018). *Interviews in qualitative research* (2nd ed). London: SAGE.

　　King et al. provide some excellent critical discussion of the foundations of interview practice and how these have evolved through time.

Platt, J. (2002). The history of the interview. In Gubrium, J. F., and Holstein, J. A. (Eds) *Handbook of interview research: Context and method* (pp. 33–54). London: SAGE.

　　This chapter documents the history of interviewing and charts this through a rich set of accounts that demonstrate how interviewing techniques have adapted and changed over time.

Rubin, H. J., and Rubin, I. S. (2011). *Qualitative interviewing: The art of hearing data* (3rd ed). London: SAGE.

　　This is, perhaps, the most widely cited book on qualitative interviewing and provides a clear and extremely detailed guide for researchers considering adopting the method.

NOTES

1　By 'messy' I mean that qualitative research, like interviews, can be uncontrollable and unpredictable (Askins and Pain, 2011).

2　Interviews are often audio recorded to aid with transcription; however, some interviewers use approaches such as note-taking, either for convenience (e.g., if conducting interviews informally or spontaneously) or if participants wish not to be recorded (see Chapter 6).

REFERENCES

Alston, W. P. (1996). *A realist conception of truth*. Ithaca: Cornell University Press.

Askins, K., and Pain, R. (2011). Contact zones: Participation, materiality, and the messiness of interaction. *Environment and Planning D: Society and Space*, *29*(5), 803–821.

Bamberg, M. (2020). Narrative analysis: An integrative approach. In Jarvinen, M, and Mik-Meyer, N. (Eds) *Qualitative analysis: Eight approaches for the social sciences* (pp. 243–264). London: SAGE.

Bingham, W. V., and Moore, B. V. (1931). *How to interview*. New York: Harper and Brothers.

Bjørnholt, Margunn, and Farstad, Gunhild R. (2014). 'Am I rambling?': On the advantages of interviewing couples together. *Qualitative Research*, *14*(1): 3–19.

Bryman, A. (1984). The debate about quantitative and qualitative research: A question of method or epistemology? *British journal of Sociology*, *35*(1), 75–92.

Bryman, A. (2016). *Social research methods*. Oxford: Oxford University Press.

Cochrane, A. (2020). Interviews. In Ward, K. (Ed) *Researching the city: A guide for students* (pp. 40–56). London: SAGE.

Denzin, N. K., and Lincoln, Y. S. (Eds) (2011). *The SAGE handbook of qualitative research*. London: SAGE.

Donalek, J. G. (2005). The interview in qualitative research. *Urologic Nursing*, *25*(2), 124–125.

Edwards, R., and Holland, J. (2013). *What is qualitative interviewing?* London: Bloomsbury Academic.

Flick, U. (2002) *An introduction to qualitative research* (2nd ed). London: SAGE.

James, N., and Busher, H. (2006). Credibility, authenticity and voice: Dilemmas in online interviewing. *Qualitative research*, *6*(3), 403–420.

Kobakhidze, M. N., Hui, J., Chui, J., and González, A. (2021). Research disruptions, new opportunities: Re-imagining qualitative interview study during the COVID-19 pandemic. *International Journal of Qualitative Methods*, *20*, 1–10.

Law, J. (2004). *After method: Mess in social science research*. London: Routledge.

McDowell, L. (1992). Doing gender: Feminism, feminists and research methods in human geography. *Transactions of the Institute of British Geographers*, *17*(4), 399–416.

Nagar-Ron, S., and Motzafi-Haller, P. (2011). 'My life? There is not much to tell': On voice, silence and agency in interviews with first-generation Mizrahi Jewish women immigrants to Israel. *Qualitative Inquiry*, *17*(7), 653–663.

Philo, C. (1992). Neglected rural geographies: A review. *Journal of Rural Studies*, *8*(2), 193–207.

Ray, L., and Sayer, A. (Eds) (1999). *Culture and economy after the cultural turn*. London: SAGE.

Reyes McGovern, E. (2019). Storytelling and mothering: A portrait of a home-grown, Mexican American teacher. *Qualitative Inquiry*, *25*(5), 482–491.

Riley, M. (2014). Interviewing fathers and sons together: Exploring the potential of joint interviews for research on family farms. *Journal of Rural Studies*, *36*, 237–246.

Robson, C. (2002). *Real world research: A resource for social scientists and practitioner-researchers*. Chichester: Wiley-Blackwell.

Rose, G. (1997). Situating knowledges: Positionality, reflexivities and other tactics. *Progress in Human Geography*, *21*(3), 30–320.

Rubin, H. J., and Rubin, I. S. (2011). *Qualitative interviewing: The art of hearing data* (3rd ed). London: SAGE.

Ryan, G. (2018). Introduction to positivism, interpretivism and critical theory. *Nurse Researcher*, *25*(4), 41–49.

Sakellariou, D., Boniface, G., and Brown, P. (2013). Using joint interviews in a narrative-based study on illness experiences. *Qualitative Health Research*, *23*(11), 1563–1570.

Savage, M. (2010). *Identities and social change in Britain since 1940: The politics of method*. Oxford: Oxford University Press.

Seidman, I. (2006). *Interviewing as qualitative research: A guide for researchers in education and the social sciences* (3rd ed). New York: Teachers College Press.

Shurmer-Smith, P. (Ed) (2002). *Doing cultural geography*. London: SAGE.

Silverman, D. (2015). *Interpreting Qualitative Data* (5th ed). London: SAGE.

Thrift, N. (2000). Introduction: Dead or alive? In Cook. I., Crouch, D., Naylor, S., and Ryan, J. R. (Eds) *Cultural turns/geographical turns: Perspectives on cultural geography* (pp. 1–6). London: Routledge.

Valentine, G. (2005). Tell me about… Using interviews as a research methodology. In Flowerdew, R. and Martin, D. (Eds) *Methods in human geography: A guide for students doing a research project*. Harlow: Pearson Education Ltd.

Wheeler, K., and Morgan Brett, B. (2021). Collecting data with interviews. *SAGE Method Space*. https://www.methodspace.com/blog/collecting-data-with-interviews

WHERE DO I BEGIN?
The basics of interview design

Chapter objectives

In this chapter, I will guide you through the practicalities of the initial stages of interview design. By reading the chapter, you should

- understand the basic foundations of interview design;
- develop knowledge of what constitutes accuracy, consistency, transparency and completeness in interview design and practice; and
- recognise the significance of sampling and the range of sampling techniques available in interview research.

INTRODUCTION

In this chapter, I consider the practicalities of interview design and will encourage you to start thinking about how to build interviewing into your research design. Often, it can be tempting to rush ahead and try to combine the 'planning' with the 'doing' in order to save time. This can inevitably lead to issues arising and the potential for time to be wasted if vital steps need revising further down the line. As with all research, a clear and well-thought-through strategy for undertaking the work that is underpinned by a strong rationale and research objectives will make any project more likely to succeed.

DOI: 10.4324/9781003292784-3

In Chapter 1, I suggested that interviews are concerned with the subjective experiences and perceptions of people, and it is this notion of subjectivity – the perceptions that emerge from a 'subject's' (or person's) individual point of view – that makes careful and thorough interview research design absolutely crucial. This is primarily because interview processes are unique to each project, meaning thought will need to go into what types of knowledge you are likely to produce, as well as how it will be produced, and the approaches you will need to take in analysing and reporting these data. Hence, because interviews are ostensibly human-to-human interactions, these will inevitably vary between each interview encounter.

The chapter is structured as follows. First, I focus on interview design to help determine the relative value of utilising an interview-based research methodology, including discussion on establishing suitable aims and objectives. Next, I will outline how to ensure rigour and responsibility are built into all stages of interview-based research, focussing specifically on the planning and practice stages of interviews. Finally, I focus on sampling strategies for establishing the participant group(s) you will include in your interview research. I will help you understand how to identify your population and then, from a suite of non-probability sampling strategies, which approach you might employ to establish your research group.

DESIGNING AN INTERVIEW-LED METHODOLOGY

Project design can be a daunting prospect, and you may have selected this book precisely to help make sense of the plethora of research methods available to you. Likewise, you may have questions about the size, shape and duration of your project and whether an interview approach might help you answer the questions that have inspired you to conduct your study. Research design is therefore one of the most important dimensions of any project. Taking time to get this right will pay dividends further down the line as you commence your research and move into the analysis and final write-up stages, so never underestimate the power of a strong research proposal. In this section, I will briefly touch on research question design and how to establish themes within research design using interviews. A key message here will be to establish clear planning strategies that can help visualise methodological plans.

What do I need to consider when designing an interview-based research methodology?

Hopefully, by this point, you will have established that qualitative interviews are going to be an appropriate research method through which to collect your data. As outlined in Chapter 1, qualitative interviews are fundamentally based on methods of analysis and explanation building that involve understanding and interpreting the complexity, detail and context of a research phenomenon. This approach contrasts with quantitative methods – such as questionnaire surveys – that seek to obtain more generalisable information about the characteristics, behaviours and attitudes of a sample population using very structured and replicable means (McLafferty, 2016). Put simply, qualitative interviews are concerned with understanding the *depth*, rather than the *breadth*, of a phenomenon.

If you are still deciding on whether interviews are right for your project, fear not; it might not be immediately obvious to you which method is most valuable to help understand your research problem, so I list a few examples of the types of research that could incorporate interviewing as a method. Projects that involve understanding people's actual experiences of a phenomenon will normally require talking to the population under investigation. Moreover, your research problem might require you to know more about the phenomenon than just *if*, *how*, *when* and *where* it exists. If your research needs to explore *why* your phenomenon might occur, then interviews are likely to provide the necessary depth that quantitative methods cannot easily achieve. Here are some hypothetical examples where interviewing could be useful:

- Examining primary school children's perspectives on veganism.
- Investigating the mobility and safety of older-aged people living in cities.
- Understanding young people's engagement with urban skateparks.
- Exploring the everyday experiences of homeless people.

All of these projects are, of course, different and will have unique outcomes depending on the social and cultural composition of your population, where they are located and when you conduct your research, but what they all have in common is a sense that talking to those affected will generate richer knowledge about the research problem than other, more quantitative or observational, research approaches.

ESTABLISHING AIMS AND OBJECTIVES

Once you have established your topic, it is important to know how you will incorporate interviews into your research design and what types of themes you might want to include in your interview approach. I will go into more detail on the actual approaches you might take to designing interview questions in Chapter 3, but first, I want to pick up on two points relating to design and anticipated outcomes.

First, it is important to consider that good research design starts with a strong, clear set of aims and objectives. These are effectively the framework upon which the rest of your research will be built, and they act as one of the key mechanisms for ensuring research is effective and meaningful. Aims and objectives will usually be generated from wider reading around a topic and feed into the types of methods used to collect data. They inform the ways in which data are analysed and written up, and form the basis of conclusions and contributions to knowledge. So, what are aims and objectives, and how do they differ from one another? All research will have a central aim (essentially, a purpose for the research). The aim is a statement of intent in that it is written in broad terms and acts as the main focus of the project. To make it achievable to investigate the aim, it is then important to break it down into manageable activities or objectives. These objectives effectively 'unpack' the aim into measurable activities and act as the steps that will be taken to achieve the aim (see Peters, 2017). In Chapter 1, I stated that qualitative research is based on the notion that reality is subjective, socially constructed and formed through multiple perspectives. Hence, research aims and objectives for interview-led research will be concerned with understanding opinions, points of view, stories and accounts, and perceptions and experiences of everyday life.

There is no blueprint for linking aims and objectives to interview design, but having a clear idea of what needs to be investigated will make it much easier to determine how this can be achieved. One way of doing this effectively is to consider what themes will be important in understanding your research problem. These are likely to emerge from your preliminary reading, and by reviewing the existing literature surrounding your topic, you will start to understand what is interesting and important, and what information might be missing (what is referred to as 'research gaps') within the literature that your research could respond to.

The following is a worked example from one of my own research projects (see Holton, 2023). Here, I conducted some research that examined young people's engagements with, and support for, coastal environments. In setting out the aims and objectives, I established a set of themes that would be appropriate to cover in my interviews. My research aim was therefore broad and was stated as follows:

> Examining how, and why, young people can meaningfully develop knowledge of, and care for, coastal environments.

Breaking down this aim, I identified three key themes that I needed to cover with my interview participants. These themes helped me establish the objectives through which I needed to investigate the overarching aim. These themes included (1) existing knowledge and understanding of coastal practices and processes, (2) everyday engagement with coastal environments, and (3) motivations for (or not) supporting coastal care. When presenting these themes as objectives they needed to be stated clearly and were discrete from one another. Here are my objectives:

This research aim will be achieved by

1. *examining how young people develop, interpret and articulate forms of coastal knowledge;*
2. *observing and discussing young people's everyday encounters with coastal environments; and*
3. *establishing if, and how, young people's citizenship helps support forms of coastal care.*

Hence, having a clear set of aims and objectives will help guide your interviews and provide a starting point from which to design an interview guide (see Chapter 3). For example, having a set of objectives means that interview themes can be set, and specific questions can be established and grouped together. This provides a framework for conducting the interviews while still allowing participants to provide more depth and context on the phenomenon and may reveal relationships and differences in opinions. For example, in my research, questions were posed to find out participants' perceptions and experiences of coastal engagements over time and how these linked with broader issues like climate change or sustainable development. Moreover, participants were asked if and how their coastal engagements affected their everyday lives, and participants were encouraged to provide examples of how this informed their daily practices. In your own research, this approach to interview design can be as structured or free-form as is necessary or desired (I will compare the different approaches to structuring an interview in Chapter 3), but knowing the parameters of your project and the key information that must be gleaned from an interview will ensure the dataset is appropriate to research problem and that time is not wasted if valuable data are missed.

Box 2.1   Activity – Designing aims and objectives

In this activity, I want to get you thinking about how you might focus a project idea into a clear set of aims and objectives.

1. Select a broad project idea that interests you – this might be a project that you are considering already, but it could also be anything that attracts your interest.
2. Once you have your project idea, try to focus it into a broader research aim. You might find this easier if you first try posing it as a question and then transform the question into a clear statement of intent. This will be the main point of the project that you are attempting to investigate.
3. Next, try identifying between two and four themes within your aim that you will need to focus on to help you split the aim into manageable chunks.
4. Finally, expand each of these themes into specific objectives – remember that these should include verbs so as to pose them as activities to be conducted during the research. These objectives can therefore be thought of as mini projects that collectively help explain the main aim.

Once you are done, reflect on the scale and achievability of your project, the different approaches that you might need to include to respond to all of the elements of the aim, and the ways in which the objectives hang together and help tell a story about your research problem.

Alongside establishing aims and objectives, understanding the anticipated outcomes of the research, i.e., working out what types of data will be produced and how these might be analysed, will help determine how to design your interview approach. It might seem strange to try and anticipate how you might deal with data at the design stage when you have not even started collecting it, but this is quite a natural part of the research design process. As with quantitative approaches that might be built around specific statistical tests, interview design requires the researcher to think carefully about how they might manage, analyse and present their data. Techniques for analysing and writing up findings will be covered in Chapters 9 and 10, but ostensibly, a researcher will need to consider at the design stage what data are likely to be produced from their interviews and their approaches to analysing these data.

For example, Chapter 9 details the approaches to coding, such as top-down 'a priori' coding systems whereby the researcher looks for pre-determined codes and themes in the data, or bottom-up approaches – such as grounded theory (Glaser and Strauss, 1967) – that allow themes to emerge from the data, rather than presupposing them. Knowing early on in the project design stages which approach might be beneficial to your analysis will mean that key themes can be identified and explored during the research process. This might be daunting, so a useful approach here can be to look at other research that has been conducted in your area that has used interviews and read about the different analytical techniques that have been employed in them. The methods sections of academic papers are an excellent starting point for this and can sometimes yield some interesting, and potentially innovate, ways of approaching interview analysis.

WHAT HAPPENS IF THINGS CHANGE?

So far, I have guided you through the preliminary planning stages of interview research design. Yet, while you may spend a long time preparing your project – designing the methods, establishing interview guides, selecting a sample and conducting the interviews themselves – research is fluid and unpredictable, meaning things can (and probably will) change. This is common, and even the most seasoned researcher will have stories of instances where research design has needed to be adapted or changed due to unforeseen circumstances. This is normal, and there is no need to panic!

For example, with all the best will in the world, your interview guide might not yield quite the types of responses you were anticipating. This is fine and may require some tweaking of the questions that are used or the delivery in which they are being asked. Say you are interviewing university students about their night-time social activities; their responses may differ depending on the point of the academic year that you speak to them (i.e., they may have less material to draw from at the beginning of the year than they might later on when they have settled in and made friends). While this could be designed into your methodology, some scenarios might not be quite as easy to spot, such as unplanned major events, which may influence how participants feel about, or report on, a phenomenon. Being intuitive to how your interview is going and responding to change are both important elements of the interview.

Moreover, you may find that your interviews generate unexpected results that differ from the research objectives stated. Again, this is fine and may require tweaking your aims and objectives to reflect the findings. For example, you might be examining people's perceptions of driverless cars in urban locations. As this technology is, to date, fairly new, it might not be possible to anticipate how people will talk about their engagements, or encounters, with it and may require some flexibility to be built into the design process to allow for unexpected responses to reveal themselves. In these instances, the data drives the project, particularly when researching new phenomena or taking novel approaches to investigating an existing issue.

BUILDING ACCURACY, CONSISTENCY, TRANSPARENCY AND COMPLETENESS INTO INTERVIEWS

Hopefully, outlining the practicalities of the initial stages of interview design has helped emphasise the professionalism that goes into interview-based research. Yet, it may still be tempting to think that interview design involves simply writing a set of questions and starting some conversations. There are, in fact, a series of other steps involved in interview research design (see Chapter 4 for a discussion of power and positionality and Chapter 5 for an outline of ethics and risk), and in this section, I want to spend some time reinforcing the importance of ensuring that credibility is designed into qualitative interviewing approaches. As I outlined in Chapter 1, debates around the scientific legitimacy of qualitative methods continue to rage on, even within the social sciences, and there is much written by qualitative researchers on the importance and validity of adopting interview approaches in research contexts (Glaser and Strauss, 1966; Cutcliffe and McKenna, 1999; Charmaz and Bryant, 2011). I therefore include this section as an opportunity to defend the subjective, partial and non-standard context of interviewing, as well as to outline the methods that qualitative researchers have adopted to ensure rigour and responsibility are embedded in interview design and practice.

In much the same way that quantitative methods must be rigorously designed, interview approaches should also be created in ways that ensure accuracy, consistency, transparency and completeness. Interviews are, by dint of their subjective and partial nature, not aiming for generalisability (i.e., they are not attempting to generate results that can be applied to a broader context), nor are they replicable in the same way as a scientific experiment. In many ways, it is the perceived lack of rigour, consistency and completeness that quantitative researchers distrust in interview-based research (Baxter and Eyles, 1997; Padgett, 2012). Needless to say, while interviews may produce data that are inconsistent, in that interviews vary in length, quality, context and content, these data must still be credible in order for the findings to be able to help researchers produce meaningful knowledge from them. To quantitative researchers, reliability means that each scientific test or experiment conducted must yield exactly the same results after repeated trials (Robson, 2002). Yet, while I have already stated in Chapter 1 that interviews are rarely exactly the same as one another, reliability also functions as part of interview-based research. However, rather than producing replicable results, the reliability associated with interviewing as a research method refers to the accuracy, consistency, transparency and completeness of data collection, analysis and presentation on behalf of the interviewer (Tatano Beck, 2021) (see Box 2.2). This type of replicability ensures that interviewers adhere to responsible approaches when designing, undertaking and analysing interview materials, which ultimately ensures interviewing is a rigorous and credible research method (Seale, 1999).

In ensuring interviews are accurate, consistent, transparent and complete, I focus here on two key phases of the interview process that relate to research rigour – planning (in relation to research design) and practice (in relation to interview administration).

Box 2.2   Building quality into interview design and practice

Roulston (2010) reflects on interview training, specifically the ways in which researchers might define and measure quality in relation to interview design, practice and analysis. Unsurprisingly, given the subjectivities associated with qualitative research, there is often no consistency in what quality *actually* means – the literature Roulston reviewed uses terms like 'validity', 'rigour', 'credibility', 'thoroughness', 'replicability' and 'representativeness', among others, to describe quality. It is this lack of consistency of what constitutes quality that detractors of qualitative interviews are most troubled by. As Roulston suggests, some have argued that interviewing does not sufficiently capture 'reality' in the sense that perceptions and experiences are not fixed in time, meaning participants might hold information back, provide misinformation or, worst of all, lie. Researchers, of course, cannot be privy to the morals and ethics of an interview participant, so it is the job of the interviewer to probe points within an encounter and then explore the relationships between the knowledge produced across the data to interpret the credibility of the accounts. Mishler (1986: 112) puts this in plain terms, claiming that the "critical issue is not the determination of one singular and absolute 'truth' but the assessment of the relative plausibility of an interpretation when compared with other specific and potentially plausible alternative interpretations". Hence, the aim here is not to standardise the interview procedure into a neat, replicable approach but instead to consider the interrelation between interview design, practice and interpretation. Roulston distils this down into four key methodological points:

1. The **appropriateness** of interviewing in relation to research questions.
2. The **quality** of the interaction itself, specifically in terms of understanding and producing knowledge.
3. The **consistency** of the research through the key phases of research design, practice and analysis.
4. The **alignment** of the methods with the aims and objectives of the research.

Planning

Planning for interviews is crucial in interview research, but due to perceptions of the free-form and 'chatty' nature of interviews, planning can sometimes be rushed or handled in a tokenistic way.

Yet planning is probably one of the interviewer's most important tools in their kit. From gaining adequate training (like reading this book!), to establishing research questions, carefully designing a research methodology, testing and piloting that methodology and seeking out the right population of participants from which to draw your sample, planning is central to the interview design process. As Founding Father of the United States of America, Benjamin Franklin, famously put it, "[B]y failing to prepare, you are preparing to fail".

The crucial first step for many novice researchers will be to make sure you know what interviews comprise, why they are appropriate to your project and how you might utilise them. This might involve attending formal interview training, reading textbooks that outline interview methods, engaging with projects, report or academic texts that have used interview methods, or talking to other researchers about their experiences. Getting a feel for how interviews have been practiced by others will help identify some of the important characteristics that may finish up in your own research design.

The other element of planning is preparation. The British Army's famous 'six p's' saying states that 'proper preparation and planning prevents poor performance'. Basically, if you understand the objectives of your project, the framework you need to work within to achieve your project, and the potential challenges and opportunities involved in conducting the research, then you are likely to be prepared enough to achieve your outcome in a professional, scholarly and sensitive way.

Practice

Practice is perhaps the more complex of these two themes, particularly as interviews involve understanding, mediating and facilitating the practices of both interview participants and interviewers, both of whom have different objectives, motivations and limitations in the research encounter. This primarily comes down to power and the various, often competing power dynamics that sit within interview situations (see Chapter 4).

Interview-based research projects will involve talking to several individuals about the research phenomenon you are investigating. This approach will result in the production of a dataset comprising a collection of separate interview transcripts. For us as researchers to be able to derive themes that respond to our research aims and objectives, we will need all of the interviews to have covered some core information and for the participants to have responded to the broader themes, questions or problems associated with the phenomenon under investigation. This does not mean standardising the interview questions into an order (if this is the case, then perhaps questionnaires are for you). Instead, we want to feel comfortable that, before each interview closes, we have covered all of the relevant information in such ways that mean something to the participant and relates to their experiences.

So, interviews may differ in length; they may have longer or shorter responses, and they may have elements that deviate from the main themes or contain other themes that were not originally considered important. This is okay. What would become problematic would be to have a set of interviews where half of the discussions are based on one theme, and the other half are based on something else entirely – perhaps if the objectives of the project have changed during the data collection. This approach, while being ethically rather dubious (see Chapter 5), would also produce a dataset that would be difficult to draw comparisons from, particularly if the interview themes and questions were completely different.

Moreover, all interviewers build ethical and moral procedures into their research design and practice in order to ensure interviews are conducted responsibly and safely. In Chapter 5 I will go into detail about the processes involved in obtaining ethical clearance for research and how to weave this into research design and practice, but in the context of being a responsible researcher, this will involve being transparent with participants about interview etiquette and conduct – both your conduct and the conduct of your participants.

In drawing these themes together, I certainly do not advocate standardising interview practice in order to create a replicable product. Doing so would render the conversational purpose of an interview almost meaningless. Instead, I support the informal, complicated and iterative processes involved in interview encounters but acknowledge that interviewers must take responsibility for ensuring quality, integrity and transparency are considered at all stages of the research design and implementation.

SAMPLING FOR INTERVIEW RESEARCH

One of the fundamental ways of ensuring rigour and responsibility in interview research is to think carefully about the population you are investigating and the type, size and shape of the sample of people from this population you will need to interview in order to generate meaningful results. In this section, I encourage you to consider who needs to be included in your research and how you intend to recruit them. I first emphasise the importance of ensuring a sample that can provide more illustrative depth to the research problem rather than one that is broad or representative. Next, I consider what a population sample looks like, how this might influence who is included in, and excluded from, the research design and what might constitute an effective sample size for a research project. Finally, I will introduce you to a range of appropriate sampling strategies for interview approaches, describe each technique as well as outline their relative strengths and weaknesses.

All research, qualitative or quantitative, must be derived from a sample of a population. When I say population, I am referring to the entire group from which you intend to draw conclusions about (Daniel, 2012). Understanding what constitutes a population is the first important step in establishing the parameters of your research design. Examples of a population might be scaled geographically (e.g., a country, region, county, or town), or derived socially (e.g., a community, an organisation or employment type), culturally (e.g., an interest group or hobby), or demographically (e.g., by gender, age, ethnicity).

Defining a population sample

Knowing early on in the project design what the parameters of your population are will help you establish the size and shape of the sample from which you need to work when conducting your interviews. As I explained in Chapter 1, qualitative research prioritises smaller samples rather than large, extensive ones in order to establish a deeper explanation of a phenomenon. Your aim, therefore, is to reach out to a sample that is representative of your population but is not so large as to become unwieldy. There is no magic number of participants to include in an interview sample, and I advise you to look at the literature in your field to see what typical samples might look like. You may, for example, have looked at the methods sections of the articles or reports you read and noted that they draw from very differently sized samples. Some projects can include interviews with significant numbers of participants, occasionally nearing 100 people, while others derive from much smaller samples (perhaps between 15 and 30 interview participants). Some research has even been derived from the experiences and perceptions of a single interview participant.

The term 'theoretical saturation' is commonly associated with interview research. This refers to the point at which no new, or additional, insights can be revealed through the data collection, and all the relevant information and findings have been identified, explored and exhausted (Bryant and Charmaz, 2007; Beitin, 2012). This is all well and good, but there is no clear definition of how saturation is reached, meaning it can be hard for new researchers to know precisely if and when saturation is reached. Moreover, sometimes the logic for adopting a sampling strategy might pragmatically boil down to time and resources (some studies might have sufficient funds to employ a researcher, or researchers, to conduct interviews, while other types of research – like student-led projects – rely on the researcher to do all the interviews themselves). There may also be instances where the population is very limited, resulting in a small, but valuable, sample. This should not be perceived as negative, and research derived from small samples, if justifiable and conducted well, can produce extremely rich and powerful knowledge.

My take-home message, therefore, is to be realistic about what you can achieve within the remit of your research, both in terms of the time and resources available to you and the ability to reach and recruit interview participants. Understanding the parameters of your sample population from the outset is therefore a crucial indicator of the feasibility of your project. For example, you cannot practically sample an entire population. There might simply be too many people to reach out to, and not everyone is going to want to be involved in your research. Likewise, if your population is too focussed, then you may find it impossible to recruit participants. To use a UK-based illustration, there might, for example, be a limited number of young male knitters living on the Isle of Wight.

Common pitfalls

Generating a workable and meaningful sample from your population is therefore a critical step in producing a set of good-quality, rigorous interviews. Hence, sampling needs to be considered very carefully when designing an interview-based project. To help emphasise the importance of rigour here, I focus on some of the ways that interview research can go wrong. There are two common pitfalls that researchers can fall into – namely, incorrect population sizes and sampling inaccuracies. Incorrect population sizes often occur when the population is poorly defined and results in sampling error or inconsistencies within the dataset (Oppong, 2013). Say, for example, you are interviewing local residents and business owners to find out if, and how, a new visitor attraction has impacted the local economy. You have defined your population as those who live and work in the vicinity of the attraction, but you have not set any parameters for who is to be included in the research. Your sample, therefore, might include people from outside the population, such as residents from neighbouring towns, commuters or tourists. While this could be accounted for in the analysis, the inclusion of those outside of the population could affect how responses are interpreted and even risk data being taken out of context or misrepresented.

Moreover, sampling inaccuracies arise if the sample is derived from only one part of a population or misses out on a section of the population. Using the same example again, say you managed to set a clear parameter for your population in the research but only interviewed men between the ages of 35 and 50 in one neighbourhood of the town. In these instances, the results may be skewed to one particular viewpoint or experience, or be misrepresentative of the true population (Goldstein, 2002). Your research may, of course, be focussing specifically on the experiences of a particular demographic, and in this case, it would be important to state this clearly, as justifiably, in the aims and objectives of the research.

Sampling strategies

There are many ways of deriving your sample, and it is worth taking time to consider which approach might be best suited to

Table 2.1 Defining probability and non-probability sampling

Probability sampling	Non-probability sampling
Sample selected using random methods	Sample selected in a non-random way
Mainly used in quantitative research	Used in both qualitative and quantitative research
Allows you to make strong statistical inferences about the population	Easier to achieve but more risk of bias

your project (see Beitin, 2012; Robson, 2002; Silverman, 2017 for detailed advice on sampling strategies). These can be collapsed into two categories: probability and non-probability sampling (see Table 2.1). Probability sampling strategies are usually used for quantitative research, such as questionnaires, that need to derive samples from very large populations. Qualitative research, like interviews, is more likely to use non-probability sampling techniques. This is often because the populations are smaller in size, might be harder to define or could be difficult to access. Researchers using interviews are likely to use techniques like *purposive, convenience* or *snowball* sampling to establish a group of suitable interview participants. These more pragmatic approaches to generating a sample might be the only way of making contact with potential participants and can sometimes take a bit more time and effort to establish the right sample needed for the study. These techniques are described as nonprobabilistic, as the samples are usually derived using non-random methods that may not relate to the entire population. An example of this might be researching with people in a vulnerable group who may be small in population, hard to reach and perhaps wary of the potential consequences of participating in academic research.

Box 2.3 Accessing 'hidden' populations

In a study of non-heterosexual women's experiences of exclusion and identity, Kath Browne (2005) explored the practicalities of sampling when attempting to access hidden populations. Browne defines hidden populations as groups with low numbers of potential participants or situations whereby the topic or context is sensitive. To develop her sample of 28 interview participants, Browne started with a form of convenience sampling, drawing upon the social networks she was already part of and then using snowball sampling to help build the rest of the sample. The strengths of adopting these sampling strategies were, first, the ability to form a sample that was appropriate to the aims and objectives of the project that might otherwise be impossible to reach using conventional advertising. Second, the use of snowballing meant that Browne was less likely to direct the composition of the group by prescribing particular identities to potential interview participants. Instead, snowballing allows participants to self-identify by providing access to a wider network of individuals. Third, as potential participants were being recommended to Browne, a degree of trust was also established that might not otherwise be possible to achieve when recruiting strangers. There were, however, limitations to this sampling approach. Fundamentally, as snowballing recruits a particular sample, it also contributes to producing and reproducing particular, or potentially partial, types of knowledge. Browne (2005: 51) refers to this as "recreating hidden populations", in that those with the most powerful voices within a group are likely to put themselves forward to be interviewed, often at the expense of those with weaker voices who remain hidden (see also Worthen's (2014) research of recruiting American participants experiencing obesity).

Non-probability sampling strategies

Quota sampling

A quota sampling technique is useful when there are specific groups within the population, and there is a need to ensure that a balanced sample is met. By setting quotas, the researcher will know if, and when, a target number of participants has been met and whether more energy needs to be put into finding participants for an under-recruiting group. For example, the location in which you want to conduct your research might have a large youth population. If you were to attempt to generate a random sample, it might capture more younger participants than those from other age groups, therefore disproportionately skewing the findings towards youth. Adopting a quota sampling technique (e.g., assigning specific age ranges to groups: 18–25, 26–35, 36–45), with upper limits to the numbers for each category, will help you identify when you have reached saturation of one category and encourage you to ensure the others reach their targets. You could achieve this using snowball sampling, for example (see the following).

Purposive sampling

Here, the researcher uses their expertise to select a sample that is most useful to the purposes of the research. This requires some background work being done into the suitability of the population to the research (e.g., a community that fits the profile of the study). In your own research, you might need to understand the perceptions and experiences of a phenomenon from a range of different stakeholders in the community. You might, therefore, develop a sample that includes, and targets, pre-determined key stakeholder individuals and groups in the community that will collectively demonstrate a broader, and potentially contrasting, set of opinions on the phenomenon.

Convenience sampling

This may sound like an unscientific approach, but convenience sampling has value when the individuals required to participate happen to be most accessible to the researcher. Examples of this might be university students or those working within an organisation or living in a community known to the researcher. You might be directly involved in the phenomenon you are investigating (e.g., it may occur where you live or work, it could be an activity that you regularly participate in or it might affect your social and/or cultural identity), and therefore, you might have a great deal of existing knowledge of the issue and have access to a relevant potential participant group from which to sample from.

Snowball sampling

This approach is widely used in qualitative research and involves asking participants currently involved in the research to help recruit new participants into the project. Snowball sampling is particularly useful when trying to reach hidden or hard to access participants (see Box 2.3). Using a snowball technique might help gain access to new recruits. Be aware that snowball sampling is likely to make the research a slow process so build in sufficient time to complete the work.

SUMMARY POINTS

- Research design is one of the most important parts of any project and must be considered carefully before any interviews are conducted.
- Establishing a clear set of aims and objectives will help determine the parameters of a research project and help define the design of interview methods.
- Qualitative research design is flexible and can be subject to change, so being responsive to issues that arise and having contingencies in place will help manage the interview process more successfully.
- For interviews to be credible and reliable, researchers must ensure that interview design, practice and analysis are accurate, consistent, transparent and complete.
- Planning is one of the most important elements of successful and responsible interview design and will make sure each encounter is professional and rigorous.
- Moreover, understanding how to interpret the 'messiness' of interview data – i.e., how each individual conversation relates to one another – will help with accuracy, consistency, transparency and completeness.
- To make research credible, it is important to define the research population carefully and then establish a clear and justifiable sampling technique through which to select participants.

WHAT TO DO NEXT?

Consider the following points:

- When piecing together your own interview design, look for examples of others who have used similar techniques and approaches as you. This can help you settle on an approach and understand what an achievable project looks like.
- Establish the key themes you need to design your interviews around.
- Be adaptable to your research and be responsive to issues, changes or unexpected findings that might emerge from the research.
- Read up on some of the literature on interview rigour and responsibility, specifically in relation to planning and practice.

- Establish the population you are researching within. Be realistic and as accurate as possible.
- When sampling, choose a strategy that is appropriate to your research objectives. Make sure that your sample is reflective of the population and develop strategies to ensure you meet your targets.

Suggested further reading

Baxter, J., and Eyles, J. (1997). Evaluating qualitative research in social geography: Establishing 'rigour' in interview analysis. *Transactions of the Institute of British Geographers*, *22*(4), 505–525.

Baxter and Eyles' critical review of rigour in interview research argues for researchers to take seriously the implications of research design. They recommend that qualitative researchers consider the criteria of credibility, transferability, dependability and confirmability for establishing rigour and to be explicit about methods in the evaluation of research design.

Beitin, B. K. (2012). Interview and sampling: How many and whom? In Gubrium, J. F., Holstein, J. A., Marvasti, A. B., and McKinney, K. D. (eds) *The SAGE handbook of interview research: The complexity of the craft.* (pp. 243–254). London: SAGE.

Beitin's chapter provides rare detail on interview sampling strategies, including the pitfalls of handling participant groups, all couched in rich and detailed research contexts.

Roulston, K. (2010). Considering quality in qualitative interviewing. *Qualitative Research*, *10*(2), 199–228.

Roulston's paper takes a critical stance on the quality of interviewing practice in order to provide recommendations for how novice researchers can ensure quality is demonstrated at all stages of the research process.

Roulston, K., DeMarrais, K., and Lewis, J. B. (2003). Learning to interview in the social sciences. *Qualitative Inquiry*, *9*(4), 643–668.

Roulston et al. draw upon researchers' real experiences of learning to interview in order to stress the importance of formal interview training.

REFERENCES

Baxter, J., and Eyles, J. (1997). Evaluating qualitative research in social geography: Establishing 'rigour' in interview analysis. *Transactions of the Institute of British Geographers, 22*(4), 505–525.

Beitin, B. K. (2012). Interview and sampling. In Gubrium, J. F., Holstein, J. A., Marvasti, A. B., and McKinney, K. D. (Eds.). *The SAGE handbook of interview research: The complexity of the craft* (pp. 243–254). London: Sage Publications.

Browne, K. (2005). Snowball sampling: Using social networks to research non-heterosexual women. *International Journal of Social Research Methodology, 8*(1), 47–60.

Bryant, A., and Charmaz, K. (Eds) (2007). *The SAGE handbook of grounded theory*. London: SAGE.

Charmaz, K., and Bryant, A. (2011). Grounded theory and credibility. In Silverman, D. (Ed) *Qualitative research* (3rd ed, pp. 291–309). London: SAGE.

Cutcliffe, J. R., and McKenna, H. P. (1999). Establishing the credibility of qualitative research findings: The plot thickens. *Journal of Advanced Nursing, 30*(2), 374–380.

Daniel, J. (2012). *Sampling essentials: Practical guidelines for making sampling choices*. London: SAGE.

Glaser, B. G., and Strauss, A. L. (1966). The purpose and credibility of qualitative research. *Nursing Research, 15*(1), 56–61.

Glaser, B. G., and Strauss, A. L. (1967). *The discovery of grounded theory: Strategies for qualitative research*. Chicago: Aldine Publishing.

Goldstein, K. (2002). Getting in the door: Sampling and completing elite interviews. *PS: Political Science & Politics, 35*(4), 669–672.

Holton, M. (2023). *Encountering coastal youth citizenship: Exploring young people's engagements with coastal environments in the UK. Final project report*. London: RGS-IBG.

McLafferty, S. L. (2016). Conducting questionnaire surveys. In Clifford, N., Cope, M., Gillespie, T. and French, S. (Eds) (2016) *Key methods in geography* (3rd ed, pp. 129–142). London: SAGE.

Mishler, E. (1986). *Research interviewing: Context and narrative*. Cambridge: Harvard University Press.

Oppong, S. H. (2013). The problem of sampling in qualitative research. *Asian Journal of Management Sciences and Education, 2*(2), 202–210.

Padgett, D. K. (2012). *Qualitative and mixed methods in public health*. London: SAGE.

Peters, K. (2017). *Your human geography dissertation: Designing, doing, delivering*. London: SAGE.

Robson, C. (2002). *Real world research: A resource for social scientists and practitioner-researchers*. Chichester: Wiley-Blackwell.

Roulston, K. (2010). Considering quality in qualitative interviewing. *Qualitative Research*, *10*(2), 199–228.

Seale, C. (1999). *The quality of qualitative research*. London: SAGE.

Silverman, D. (2017). *Doing qualitative research* (5th ed). London: SAGE.

Tatano Beck, C. (2021). *Introduction to phenomenology: Focus on methodology*. London: SAGE.

Worthen, M. G. F. (2014). An invitation to use Craigslist ads to recruit respondents from stigmatized groups for qualitative interviews. *Qualitative Research*, *14*(3), 371–383.

STRUCTURING AN INTERVIEW
Designing interview questions

Chapter objectives

This chapter explores a range of approaches to structuring and designing effective interview questions. By reading the chapter, you should

- develop knowledge of a range of interview design approaches,
- comprehend methods for designing clear and appropriate interview questions,
- understand how to produce and use interview guides, and
- appreciate the importance of practicing and piloting interviews.

INTRODUCTION

In this chapter, I will move from the broader design phase of an interview, as outlined in Chapter 2, into the specifics of designing effective interview questions. Here, I will take time to provide guidance on the basics of interview structure and advise you on the strategies you can adopt when designing your own interview questions. As I will explain throughout the chapter, question design can be tricky, and it is important to get this stage of the research design right so as to ensure the data produced is effective at responding to your research aims and objectives (Brinkmann and Kvale, 2018). For example, being clear with yourself and your participants on the interview approach you want to take will make the conversation flow more naturally, as will having some form of interview guide that scaffolds the themes and questions you want to talk about. Within this guide, preparing questions that encourage fuller, more explanatory responses will help make the analysis process more interesting and fulfilling, as there will be more evidence for you to draw upon. These approaches might not be familiar to you as yet, so practicing your interviews using a piloting scheme is likely to help you understand how future participants might respond to questions and encourage you to make changes if there are weaknesses in the interview design.

In structuring the chapter, first, I examine three key interview approaches – structured, semi-structured and unstructured interviewing – and consider when these are most appropriately administered. Next, I outline some strategies for designing interview questions and the ways in which closed (or factual) and open (or probing) questions can be built into an interview structure. Third, I explore interview guides as a tool for managing the interview process that builds transparency and accuracy into research design, as well as encouraging a greater ability for thematic replication across the set of interviews. Finally, I consider the importance of testing and piloting within interview design to ensure all aspects of the approach work correctly. Crucially, rather than operating as a 'how-to' guide, I have structured this chapter so as to encourage you to think carefully about what you want 'your' interview approach to achieve, using examples from a range of research projects.

STRUCTURED, SEMI-STRUCTURED OR UNSTRUCTURED INTERVIEWS?

In this section, I explore the various types of interviewing approaches available to researchers. The reason for including this in the chapter is to convey the importance of understanding interviews as individual encounters, of which no two will be exactly the same or follow precisely the same patterns of responses. This may fill you with dread, particularly if you consider data to be rigid and scientifically replicable. However, it is precisely this flexibility and difference that makes qualitative interviewing important. I intentionally labour the point that interviews, by dint of their name, almost always deal directly with humans on a one-to-one basis, and therefore each conversation will have to account for a participant's personality, experiences and backgrounds, as well as their morals, beliefs and ethics. Moreover, the interviewer is also human, and as will be outlined in

Chapter 4, recognising who *you* are as a researcher, the relationship *you* have with your participants and how this relationship affects the type and quality of knowledge being produced is hugely important. This section will therefore examine the three ways that researchers might consider designing an interview approach. These approaches are categorised into three 'types' of interview – structured, semi-structured and unstructured. In each instance, I will explore the approach and provide examples of how they have been used in social science research. An overview of the three approaches can be found in Table 3.1.

Table 3.1 Types of qualitative interviews

Interview approach	Key features
Structured interviews	Mainly used in survey research. Tightly defined interview – sometimes referred to as an 'oral questionnaire'. Can be quick to implement and generate large samples. Led primarily by the interviewer.
Semi-structured interviews	Researcher creates an 'interview guide' containing a list of topics to be covered. Flexibility in how participants respond and in how questions are ordered. An underlying intention remains that some, or most, questions are asked consistently. Led by the interviewer but with opportunities for flexibility.
Unstructured interviews	A brief list of potential topics is generated. Interview direction is completely open, with longer-form responses from participants. Useful for new avenues of research (e.g., topics that have not been researched before). Led primarily by the interviewee.

Structured interviewing

The first type of interview is the structured interview approach. This is a very formalised way of designing and conducting interviews that follows a strict, formulaic script and does not allow for any deviation (Dunn, 2016). These interview types are often highly organised and can sometimes be defined as 'oral questionnaires' or 'street interviews' (Strebel, 2014), as they can be conducted outside and rely on short responses that can be written down quickly and generate large samples of responses. Structured interview approaches are mainly used in survey research and can comprise a mixture of closed and open questions that do not require much opportunity for nuance or deviation. That said, while questions are fixed in structured interviews, they still allow participants the freedom to answer openly, as opposed to questionnaires that primarily use closed questions that limit participants to responding to prescribed lists of options to choose from[1] (Clark et al., 2019).

You may have participated in a structured interview in your local high street, outside a sporting event or in a supermarket, and we often link these types of methods to market research, whereby an organisation is canvassing opinion. While this approach is ostensibly fine for scholarly research, it is important to design a method that uses this structure to its best advantage that could not otherwise be achieved better using a questionnaire. For example, your research project may be concerned with understanding how sports fans might experience and express their 'fandom' during their favourite sporting event. Conducting short, structured interviews at the stadium, either before or after the event, might help you be 'in the moment' with your participants (and not over-burden their time) in ways that a questionnaire or a longer interview might miss if conducted at a different time or place.

Structured interview designs can also be useful for novice interviewers who may be anxious about doing interviews for the first time or lack confidence in leading the interview process. I remember vividly the first interviews I carried out for my undergraduate dissertation, where I had prepared a list of 30 questions to ask my participants and doggedly stuck to the script, even though I knew there were things that my participants were saying that I wanted to know more about. While the data generated were interesting and related to my research objectives, I quickly realised that my structured approach left little room for my participants to offer their own reflections on the themes I was asking them about. At this point, I want to stress that adopting a structured approach is by no means wrong. It does, however, limit the extent to which a partici-pant can express their own voice and may require the interviewer to step back a little to consider whether this approach might lead to narrow or very specific knowledge being produced that might not take into consideration the nuanced experiences and perceptions of the participant group (Winchester and Rofe, 2016).

Semi-structured interviewing

The final caveat from the previous section segues neatly into the justification for more semi-structured interview approaches. This is by far the most popularly used interview technique, as it allows for some structure to be retained but provides participants (and researchers alike) the opportunity to deviate from the questions posed, or even approach them in a different order for each interview. Semi-structured interviews contrast from structured approaches because they are thematically driven (Roulston and Choi, 2018). This means that the interview has a pre-determined framework of topics that need to be covered but allows for deviation during a research encounter (O-Reilly and Dogra, 2017). This might involve an interview guide that uses themes as headings and perhaps lists key questions that need to be asked and subsidiary topics, questions or prompts that might be necessary should they not naturally come up in conversation.

Semi-structured interviews can, therefore, be extremely valu-able in ensuring you cover the requisite themes needed to answer your research aims and objectives, as they provide the chance for the researcher to step back and let participants talk a bit more freely about their experiences. Sometimes this might mean that you lin-ger on a theme for some time while a participant talks at length about it. In other places, it might feel more comfortable to move on or revisit a theme later if a participant is happy to do so, or perhaps starts speaking to a theme earlier on in the interview than you have scheduled to talk about. This obviously requires the inter-viewer to concentrate carefully on the encounter and to have a clear handle on what has been covered and what remains outstanding. A semi-structured approach can therefore take some perfecting but is a valuable skill once learned. Crucially, this means that the transcripts that are produced from each semi-structured interview are likely to vary greatly from one another – very different from a more structured approach that would yield responses in a specific order. Yet, if approached carefully, this will not matter too much, as the themes will ultimately be covered in every instance.

Unstructured interviewing

Occupying the opposite end of the spectrum to structured approaches are unstructured interviews. This style of interviewing differs completely from structured approaches in that they are entirely freeform and open-ended and contain no pre-determined framework or questions (O-Reilly and Dogra, 2017). Unstructured interviews are ostensibly led by the conversation between the interviewer and the interview participant and have minimal pre-determined direction from the interviewer. They are likely to contain a brief list of broad topics that need to be covered but the direction of how these are discussed is left completely open. This type of approach might be instantly off-putting to a novice researcher, particularly as having no structure can be hard to implement or even feel unscientific, yet unstructured approaches are likely to produce some of the most natural responses from participants than set questions.

Unstructured approaches might be useful when researching a brand-new phenomenon in which the researcher may not know anything about how the participants might engage with or perceive the situation, or issue, they are experiencing (Billups, 2019). Research into people's engagement with new digital technologies, for example, may require interviewers to encourage participants to open up and talk about their experiences, such as how they use the technology, examples of how it has improved or hindered their everyday lives or whether they believe it will become part of their common routines. Starting the interview process in an unstructured way may therefore reveal unexpected or new themes that can then be incorporated into further interviews and can help strengthen the interview process as a 'design in progress' that can morph and adapt over the duration of the research.

In my most recent research, I adopted an unstructured approach for the first time in my career. My interviews with young people engaging with coastal and marine environments needed a lot of flexibility designed into them (see Holton, 2023). The members of the participant group were all of a similar age, but their encounters with the sea were varied (some swam for leisure, while others operated ferries for work), and their motivations for engaging with the sea were predicated on environmental concerns, their mental health or their family histories. Moreover, their backgrounds affected their engagements, with some participants able to experience the sea every day or using expensive equipment, while others rarely or never had opportunities to engage due to costs or location. Hence, while a structured or semi-structured approach could have identified these differences, taking an unstructured approach that allowed participants to tell me about their perceptions and experiences in their own ways very much drew out the nuance of their encounters and provided a richer set of interview data.

INTERVIEW QUESTION DESIGN

Once you have settled on your interview approach you will then need to consider the types of questions you will ask during the interview encounter. Right at the beginning of this book, I stated that an interview is a conversation with a purpose. It is precisely this conversation that will produce the types of knowledge necessary to respond to the research aims and objectives you propose. This conversational approach may sound rather simplistic, but for responses to be free, open and honest, the interview questions must be posed clearly, sensitively and in ways that allow participants to respond in their own way.

Robson (2002) proposes three themes for designing interview questions that are based on (1) obtaining facts; (2) understanding participants' behaviours, experiences and perceptions; and (3) examining people's beliefs or attitudes. Factual information is usually fairly simple to acquire from participants (e.g., questions that ask a participant's age, location, employment status), and it may be important to gather specific information about the phenomenon you are investigating that can later be compared across the dataset. A word of caution, though, is to avoid posing too many factual questions as closed questions. A closed question that only elicits a 'yes/no' response is unlikely to yield much meaningful data at the analysis stage. You therefore may need to have a follow-up question, or set of questions, that delve deeper into why the participant responded in the way they did (see Box 3.1). It is also important to limit the number of factual questions in an interview so as not to fall into the rhythm of interrogation and consider carefully where they are placed in the structure of your interview.

Box 3.1   Types of factual questions used in interviews

There may be certain types of information that you need to elicit from your interview participants that come in the form of closed questions. These are likely to be contextual questions in that they are informative, but you will need to follow them up with open questions to help propel the conversation. The following are some examples of the types of closed questions that might be asked in an interview:

- "How long have you lived in this town?"
- "Which school did you attend?"
- "Do you use public transport to commute to work?"
- "What is your job role?"

Note that these types of closed questions are quite similar to those used in questionnaire surveys that provide structured ways of responding. Hence, as Robson (2002) suggests, closed questions may well fix the structure of an interview by preventing participants from answering freely, so limit their usage.

Questions about behaviours, experiences and perceptions are also fairly straightforward to design into an interview, particularly as these are likely to be intrinsically tied to the research problem you are investigating (see Box 3.2). As such, open questions – what we may also call 'probing questions' – need to be posed carefully to ensure participants provide longer, more generous responses than perhaps to factual questions. Valentine (2005) suggests that presenting probing questions using the phrase 'tell me about…' encourages more flexibility in how a participant responds to a question while still retaining the focus of the interview. Asking participants to tell you about their daily commute to work, their retail behaviours or favourite sporting activity can only really elicit longer responses. It would be unnatural to respond with anything shorter unless the participant

> **Box 3.2 Interview question checklist**
>
> I have adapted from Robson (2002) a simple checklist that can help determine the value of the questions you might include in your interview.
>
> • Is the question flexible? Does it allow the participant to answer it in their own way?
> • Does the question encourage longer, fuller responses?
> • Is the question inclusive? Will the participant feel comfortable answering the question?
> • Does the question invite an honest response? Can the integrity of the response be tested at all by posing other questions?
> • Is the question open to additional or unexpected responses?

flatly refused to answer your question. Remember though, as with the factual questions, to vary the types of probing questions posed. Questions like, "How would you describe your daily commute?" or "Why did you decide to take up netball as a child?" can be just as effective at encouraging deeper responses. Just be aware to avoid posing leading questions that impose certain responses from participants, such as, "Should the city authorities spend even more tax money trying to keep the streets clean and tidy?" Leading questions are the antithesis of probing questions in that they narrow the focus of the interview and limit the scope to investigate a range of potential responses from participants.

Questions relating to beliefs and attitudes are usually the most challenging to get right in an interview. Too pointed and a participant may become offended. Too vague and the participant might not understand what is being asked of them. It is also unlikely that a participant's beliefs or attitudes can be captured in a single 'tell me about…' style question. Indeed, such issues can take up a significant proportion of the interview and may require the interviewer to ask additional follow-up probing questions that test or challenge participants' beliefs and attitudes. Many researchers can probably recall instances where they have felt a particular interview participant was telling them what they thought the researcher wanted to hear, and these responses can come across as rather bland and descriptive.

In these instances, asking for examples of their perceptions and/ or experiences of the phenomenon you are investigating can encourage participants to open up and elaborate on their responses. Remember, though, that the key word here is 'encourage', and it is very important not to let the interview slip into a form of inter-rogation. Probing too hard, i.e., repeating a question over and over again to get a response, is a hostile practice, and interview partici-pants are likely to get upset if they feel they are being pressurised into responding to something. As an interviewer, you are aiming to record your participants' experiences, perceptions and interpreta-tions rather than seeking out 'the truth'. Hence, it is important to accept the responses given and look for patterns and discrepancies across the range of interviews you have (see Chapter 9).

The final part of question design pertains to the sequencing of the questions that are being posed. As stated earlier in the inter-view approaches section, the conversational nature of interviews means they are rarely sequenced in exactly the same way for every participant, and the human quality of interviews will undoubtedly encourage deviation from the script – particularly when utilising semi-structured or unstructured approaches. There are some com-mon strategies that can be built into interview design, though, that help manage the flow of the conversation and ensure the interviews are rigorous, complete and conducted responsibly.

All interviews require a clear and effective introduction that briefly and simply explains the purpose of the research and the par-ticipant's role within it, reminds the interview participant of the ethical considerations associated with the interview (such as confi-dentiality), requests permission to record the interview and signposts the interview themes. After the introduction, a set of factual, non-threatening 'ice-breaker' questions can be valuable in settling both the interview participant and the interviewer into the rhythm of the interview. These might involve asking simple, general questions similar to the closed questions outlined earlier in this section. The main body of the interview will usually be divided according to the research themes and will contain a range of different open questions designed to elicit longer responses (see the section on open ques-tions for ideas). Setting aside time at the end of the interview to cover any missed themes or elaborate on responses will help wind down the interview and bring it to a natural close.

Box 3.3   Activity – Designing and testing a set of interview questions

Have a go at designing a set of interview questions. On your own, design a ten-minute 'semi-structured interview' on the topic, "What attracted you to take up XXXX activity?" (You can choose the activity.) You will need to include the following questions:

- Two 'ice-breaker' questions
- Two probing questions
- Two follow-on, or backup, questions

Remember that probing questions will need to be open, so starting with 'tell me about...' or 'why...' will encourage your participants to provide longer responses.

Next, with a friend or colleague, conduct your interview. During the interview, the interviewer and interview participant should reflect on the following:

- **Interviewer**: make note of how the interview participant *responds* to the questioning.
- **Interview participant**: make note of how the interviewer *poses* the questions.

After the interview, discuss with each other:

- Did the interview flow logically?

If not, would you word the questions differently and why?

PRODUCING AND USING INTERVIEW GUIDES

One method through which interviewers can help build structure into an encounter is by producing an interview guide. This section will introduce you to what an interview guide entails, the types of 'ingredients' involved in producing an interview guide and the value of using one in structuring and supporting the interview process – particularly for novice researchers. I will caveat this section by saying that not everyone uses an interview guide, and what I am proposing here may sound very formal and overly prescriptive. This is somewhat intentional at this stage of attending to the basics of interview design, as I want to provide a set of considerations from which you can proceed with your own research. So, what I suggest below might work for you; however, you may choose to set out a rigid format of questions, produce some loosely defined but consistent themes, or take a completely unstructured approach – I have certainly adopted all three in my own research!

An interview guide should be used as a device that establishes and guides the focus of an interview but should not prescribe or overly direct the process. Interview guides may therefore be produced as an outcome of interview piloting and should be considered a dynamic, rather than rigid, tool. Certain themes may reveal themselves during the interview process that you might want to ensure are covered elsewhere. Your interview guide should therefore be considered a reflective instrument that helps you stay focused on the task at hand.

There is no set format for producing interview guides, and you may want to talk with your teacher or advisor – or someone else familiar with interview research – about their experiences of using them or not. In Figure 3.1, I outline some potentially useful components that can help with structuring a guide, and you may consider adopting these depending on your skill and confidence in running an interview (see Box 3.4 for advice on constructing your own interview guide). Thematic headings are valuable in ensuring the interview is bounded and relates clearly to the issues, problems or topics identified in the aims and objectives. A clear set of instructions and/or interviewer prompts will guide you through the process and help get you back on track should you get lost in the interview process. Starting an interview can be daunting and feel awkward, so having some simple 'ice-breaker' questions will help interview participants settle into the conversation. These might involve preliminary or warm-up questions that outline the theme and collect general information. The next layer of questions will include some form of probing questions. These may be highly structured or more thematic prompts, but their inclusion in the guide means you will be able to delve deeper into the topic and encourage fuller answers or more specific responses. In addition to the core questions you need to ask, you might also have further probing questions that act as backup questions if necessary. This can be valuable if participants perhaps provide more descriptive responses.

Explain the interview process to the interviewee and ensure informed consent is obtained. Check that the interview can be recorded and that the interviewee is comfortable.

Ice-breaker questions:

What type of farming do you do?

Tell me about your history of farming? Have you got a family history of farming?

Everyday practices on the farm:

- An exploration of day to day behaviours, activities and routines.
- Where these are conducted, who with and how much of the day they take up.
- What are the emotional dimensions of this?:
  - How does the participant feel when conducting these tasks?
  - What are they thinking about?
  - Do they enjoy solitary tasks or look forward to more social ones?
  - Has this changed over time and what are the implications for this?

Social networks (physical as well as online):

- Exploring how connected – or not – the participant is to others:
  - On the farm,
  - In the local community;
  - Beyond the community – e.g. social media.
- How important are face to face interactions with the above?
- How are they communicating and making contact with others?
- Are there differences in how they perceive face to face and online connections?
- Has the perception of farming communities changed or remained static over the years, and why might that be?

The blurring of home and working life:

- Start to explore family life in relation to the farm here.
- How long has the farm belonged to the participant or their family?
- Tell us about the history of the farm?
- The role of the family on the farm?

Farming identities and the image of the farmer:

- What does the participant perceive the image of the farmer to be?
  - Physically
  - Emotionally
- Is this realistic?
- Is this problematic?

Figure 3.1 An example of an interview guide

Box 3.4   Activity – Designing an interview guide

I want you to design an interview guide for a research project entitled: "Investigating the Decline of Urban Retail Centres". We are interested in how the rise in online retailing has led to traditional urban retail centres becoming increasingly empty spaces. We want to find out through interviews with shoppers and business owners whether the decline in footfall has led to a rise in anti-social behaviour and whether urban, state-led initiatives, such as pop-up shops, new service provisions and cultural events, might help reinvigorate the city centre.

To produce your interview guide, consider the following steps:

1. Establish three to four themes that you want to cover as part of the interviews.
2. Write out a short set of prompts that will help guide you through the interview.
3. List three preliminary ice-breaker questions that can help you gain some general information from your participants.
4. For each theme, write out three probing questions (if you are feeling confident and want to take a more unstructured approach, then you could list these as sub-themes instead).
5. List a further follow-up question for each probing question/sub-theme. These could be placed in parentheses after the question.

Once you have completed your guide, consider the following questions:

- Is the interview guide sequenced logically? Do the questions or themes naturally flow from one another?
- Do your questions capture all elements of the research problem identified?
- Are the themes discrete from one another (i.e., can you contain your questions within a single theme without drifting into another)?
- Is there flexibility in the guide should an interview participant go 'off topic'?

Using your interview guide

The interview guide should be treated as a set of prompts for you as the interviewer. You would not normally provide the guide to participants in advance, but instead, at the start of the interview, you might indicate the general themes or areas that will be covered in the encounter. Note that in the example in Figure 3.1, I use a mixture of prompts and questions. There may be some things that you deem important to cover, and hence, a specific question might be appropriate to include. You may also find that some questions naturally require further probing; hence, if a participant answers a question a little too succinctly, then you might want an additional set of questions or prompts that will help expand on their response. This might be as simple as asking questions like "*What do you mean by XXX?*" or "*Can you give me an example of XXX?*"

PRACTICE MAKES PERFECT – PILOTING INTERVIEWS

An interview pilot is, as the name suggests, an opportunity to test out the interview format, themes and questions before conducting the research (Bloor and Wood, 2006). Piloting is an important yet often overlooked aspect of research design. It is tempting to want to rush excitedly into the 'doing' phase of research, and I have seen many instances where novice researchers commence interviewing very quickly only to find issues with the research design or delivery that compromise the integrity of the research. You only get one opportunity to work with your participants, and it is vital that you get this right the first time. Piloting the interview design is therefore an important step in reviewing, adapting and modifying the interview structure.

Pilot interviews may have different guises and motivations, depending on their intention. You might be unfamiliar with interviewing altogether or have experience of interviewing but perhaps want to explore how your questions might come across or be perceived. Moreover, you may be conducting research on a brand-new phenomenon and therefore need to explore what types of themes could be valuable to discuss with potential participants. Likewise, your interview design might contain additional methods that use technological equipment or require participants to carry out activities that may require you to test or trial before you start collecting data (Blake, 2015). Whatever your situation, piloting will ensure you are fully prepared for the real thing and convey a sense of professionalism to the interview procedure.

Ordinarily, any information gathered from the piloting stage should not be included in the finished research. To ensure the pilot study is successful, it would ideally be conducted with a group that is not connected to the research participant sample. The best interview pilots are those that have two stages. The first stage is to test the interview itself, following the schedule of activities, themes and questions as closely as possible. The second stage is reflective and should involve asking participants for feedback on the interview process. This reflective exercise might focus on the themes, lines of question, your interview approach and the duration and pace, among other things. You might collect this information as part of an informal chat at the end of the encounter or in a post-interview survey if conducting a larger pilot. However this is done, the feedback gained from this can be used to refine the methods, focus the questioning or themes and polish your interviewing technique.

SUMMARY POINTS

- Understanding what types of themes are likely to be generated from a set of interviews will help with preparing an effective interview guide.
- Interviews are designed according to three approaches – structured, semi-structured and unstructured interviewing.
- Interview questions are usually based upon three themes: obtaining facts, understanding behaviour and examining beliefs or attitudes.
- Good interview questions should elicit longer, fuller responses from participants; therefore, open questions, such as 'tell me about…' questions, will ensure the interview is more of a conversation.
- Interviewers should be prepared to be flexible in the sequencing of questions, particularly if participants go 'off topic'.
- Interview guides are a valuable way of establishing the necessary 'ingredients' of your interviews and how to manage conversations to confirm that these are covered effectively.
- Piloting, or practicing, your interviews will help identify errors, problems or gaps in the research design and help you and your participants feel comfortable during each encounter.

WHAT TO DO NEXT?

Consider the following:

- Think about whether a structured, semi-structured or unstructured approach is appropriate for your research.
- Consider carefully the types of open questions you might use in your interviews.
- Establish an interview guide or plan that will help you manage each encounter. This need not be overly prescriptive but can be valuable in setting parameters and understanding how best to address the research aims and objectives.
- Practice your interview technique to make sure you are able to manage all of the elements of the interview encounter.

Suggested further reading

Bloor, M., and Wood, F. (2006). *Keywords in qualitative methods*. London: SAGE.

Bloor and Wood supply a clear and concise set of simple approaches to piloting interviews that new researchers should find reassuring.

Roberts, R. E. (2020). Qualitative interview questions: guidance for novice researchers. *Qualitative Report, 25*(9), 3185–3203.

Roberts' paper provides excellent advice on developing an effective interview guide based on the effective structuring and ordering of questions.

Roulston, K., and Choi, M. (2018). Qualitative interviews. In Flick, U. (Ed) *The SAGE Handbook of Qualitative Data Collection* (pp. 233–249). London: SAGE.

This chapter provides clear advice on the approaches to interview design, including preparatory techniques and applications that help support future interview practice.

Wang, J., and Yan, Y. (2016). The interview question. In Gubrium, J. F., Holstein, J. A., Marvasti, A. B., and McKinney, K. D. (Eds) *The SAGE handbook of interview research: The complexity of the craft* (pp. 231–242). London: SAGE.

This chapter explores the etymology of the word 'question' and proposes some interesting approaches to interview question design that can help new interviewers construct appropriate questions within their research design.

NOTE

1   Questionnaires can also contain open-ended questions, but these are usually used sparingly to ensure the questionnaire does not become a time burden to complete.

REFERENCES

Billups, F. D. (2019). *Qualitative data collection tools: Design, development, and applications*. London: SAGE.

Blake, M. (2015). *Cognitive interviewing practice*. London: SAGE.

Bloor, M., and Wood, F. (2006). *Keywords in qualitative methods*. London: SAGE.

Brinkmann, S., and Kvale, S. (2018). *Doing interviews*. London: SAGE.

Clark, T., Foster, L. and Bryman, A. (2019). *How to do your social research project or dissertation*. London: Oxford University Press.

Dunn, K. (2016). Interviewing. In Hay, I. (Ed) *Qualitative research methods in human geography* (4th ed., pp. 149–188). Ontario: Oxford University Press.

Holton, M. (2023). *Encountering coastal youth citizenship: Exploring young people's engagements with coastal environments in the UK. Final project report*. London: RGS-IBG.

O-Reilly, M., and Dogra, N. (2017). *Interviewing children and young people for research*. London: SAGE.

Robson, C. (2002). *Real world research: A resource for social scientists and practitioner-researchers*. Chichester: Wiley-Blackwell.

Roulston, K., and Choi, M. (2018). Qualitative Interviews. In Flick, U. (Ed) *The SAGE handbook of qualitative data collection* (pp. 233–245). London: SAGE.

Strebel, I. (2014). Re-specifying geographical quantification: Problems of order in street interviews. *Transactions of the Institute of British Geographers*, *39*(2), 278–290.

Valentine, G. (2005). Tell me about… Using interviews as a research methodology. In Flowerdew, R. and Martin, D. (Eds) *Methods in human geography: A guide for students doing a research project*. Harlow: Pearson Education Ltd.

Winchester, H. P. M., and Rofe, M. W. (2016). Qualitative research and its place in human geography. In Hay, I. (Ed) *Qualitative research methods in human geography* (4th ed., pp. 3–28). Ontario: Oxford University Press.

MAKING SENSE OF POSITIONALITY AND POWER IN INTERVIEWS

Chapter objectives

This chapter examines the important role of positionality and power in producing (and suppressing) the knowledge that comes from interviewing participants. By reading the chapter, you should

- develop awareness of what is meant by positionality and how this might affect the design, practice and analysis of interviews;
- be aware of the role of power and authority in influencing the type and quality of knowledge produced in an interview;
- recognise the differences between 'insider' and 'outsider' positionalities; and
- understand how to reflect on positionality and power during the research process using a research diary.

INTRODUCTION – WHY SHOULD I CARE ABOUT POSITIONALITY AND POWER WHEN INTERVIEWING?

In Chapters 2 and 3, I discussed the planning phase of qualitative interviewing to show you how, through careful design, testing and application, interviews can be a vital research tool. In this chapter, I want to step back slightly to return to the debates discussed in Chapter 1 on subjectivity and the role of the human in interview research. My intention here is not to turn this into a philosophical debate but to remind you that subjective, qualitative research refers to research that focuses on the experiences and perceptions of individuals and examines these through the observations and interpretations

DOI: 10.4324/9781003292784-5

of the researcher. One of the key criticisms of interview methods (and qualitative methods more generally) is that they cannot account for, or control, human subjectivity. While quantitative experiments can be administered in controlled laboratories, using specific methods that are not (usually) influenced by the researcher, qualitative methods – especially interviews – are messy (in that each interview is not conducted in exactly the same way); hinge on people's moods, emotions and circumstances; and are influenced by the interaction between the interviewer and interview participant. This last point about the interview interaction – or what I will refer to in this chapter as positionality and power – is central to this chapter, and I will explain here how and why positionality and power can inject vital energy into an interview encounter or be destructive or counterproductive if not handled correctly. I have therefore intentionally included a question in the title of this introduction: "*Why should I care about positionality and power when interviewing?*" You might think that a conversation is surely *just that*, a conversation, and that you would not possibly let anything imbalance that. Yet, positionality and power are probably the central and often most overlooked drivers of all interview encounters. Positionality and power therefore both play significant roles in the methodological design, sampling, recruitment, implementation and analysis stages of the interview process.

The chapter is structured into four sections. I begin by introducing you to the notion of positionality and what effect this can have in the design, implementation and analysis of qualitative interviews. Next, I examine the role of power and the challenges of managing the 'interviewer'/'interview participant' roles during interview encounters. Third, I consider the role of 'insider' and 'outsider' positionalities when conducting research. Finally, I extol the value of using research diaries to help acknowledge positionality and aid reflexivity during the research process.

WHO AM I, AND WHO ARE MY PARTICIPANTS?

Let us start this discussion with positionality. Since the 1990s, social researchers have fought hard to contest the notion of knowledge being value-free and the product of a single truth, and qualitative methodologists have rallied for greater attention to be given to how knowledge is produced, for whom and under what circumstances (Shurmer-Smith, 2002). This centres on the intrinsic role of the researcher as a human being who designs, conducts and analyses data, and, certainly in terms of interviewing, how the backgrounds and motivations of the interviewer and the interview participant relate to one another in specific and unique ways when producing knowledge. We refer to this as *positionality*: effectively the ways in which our identities, social position and culture might shape who we are, what we believe in and how we act (Kezar, 2002). Our positionality affects our own access to aspects of society but can also be read in relation to others' positions too, meaning my positionality is likely to lead *me* to see the world differently from *you*.

In research contexts, positionality is often viewed as problematic to those seeking to understand the world from an objective standpoint that presupposes research to produce a single 'truth' about a problem (remember that subjective, qualitative researchers recognise that individuals or cultures create, or interpret, their own version(s) of 'reality', meaning there is no singular truth about a problem, rather a variety of representations of how a problem is understood). For example, might your position cloud your judgement when researching a phenomenon? Or could you be biased towards or against certain opinions that come from your research participants? Kim England (1994: 248) writes that "the researcher is a visible and integral part of the research setting", meaning the act of doing qualitative research makes us actively involved in the type and quality of knowledge that is produced. As interviewers, our identities are therefore unique to us and shape our worldview. For example, my position as a white, middle-class, middle-aged, cis-gender man living in the United Kingdom might provide me privileged access to certain forms of knowledge or experiences but limit or prevent me from having knowledge or experiences of other things.

We might, therefore, consider our positionalities to consist of a variety of factors. Some of these might be visual and therefore easier to identify, such as our age, gender, ethnicity, race, (dis)ability, among others. Yet, our positionalities are also formed through a myriad of hidden social factors, including our beliefs, class position, education, ethics, morals, culture and customs (Secules et al., 2021). All these factors intersect to produce our own specific worldviews, and we may not be aware of how our own positionalities as researchers might align with the positionalities of our participants during an interview encounter, and importantly, how we, and our participants, navigate between seemingly apparent and assumed identities during research encounters.

This might, at first, all seem very unscientific, yet recognising and understanding the role of, and implications for, positionality makes it a powerful implement in the interviewer's toolkit. To acknowledge your own positionality is therefore vital in ensuring you, as a researcher, are self-reflexive (i.e., self-aware) of how your values and experiences might influence both the way an interview is conducted and the types of knowledge that can be produced through the interview. In Box 4.1, I have outlined some of the ways that positionality might be considered before, during and after an interview encounter.

Box 4.1 Understanding the role of positionality in interview research

Positionality plays a role in all aspects of an interview encounter. The following are some questions you might consider asking of yourself as you plan, implement and reflect on your interviews.

Before starting

- What are the power dynamics between the researcher and the researched?
- Are you an insider or an outsider?
- What ethical issues do you have to consider?

During

- How are the participants reacting to my presence in the interview?
- Am I reproducing stereotypes?
- Am I assuming the answers they are giving me?

After

- Did my position/stance/opinion change during the research?
- How did different social factors affect interactions?
- How did my participants perceive me?

A FEW WORDS ABOUT POWER

While I began this chapter by questioning, "Who am I?", in this section, I pose an altogether different question: "Who are *we*?" so as to probe deeper into the dialogic qualities of an interview encounter. Our positionalities will influence how we manage our identities in relation to our participants during an interview. Yet, there is more to it than that, and any encounter that involves purposeful communication between two (or more) people will be influenced in one way or another by power (Pini, 2005; Cochrane, 1998). Power is, in itself, an interesting word. Dictionary definitions of power often refer to nouns like 'control' and 'strength'; or terms like 'official rights', which all denote power to be hierarchical; or for power to be asserted by one person over another, i.e., for one person to have power over someone else with less or no power. Yet, power can also be something more personal or embodied (essentially something we carry inside us). Other nouns, like 'ability', 'capability' and 'capacity' imply power to be something that drives us or motivates us, or provides us with the ability to overcome a problem or complete a task or activity. We may even consider this way of thinking about power as oppositional to the aforementioned hierarchical terms, whereby we actively share our power with others rather than inflict it on them. Whichever way we consider it, power is highly significant in how we communicate with one another. Power both produces knowledge and is a tool that can suppress which knowledge is produced and how it is disseminated (Adriansen and Madsen, 2014).

Let us think about this practically. Consider a situation where you have felt powerless, like during a job interview or in a meeting with someone in authority. How did you feel? Were you in control of the direction of the conversation? Were you able to freely ask your own questions, or were you only expected to provide responses? Were you compelled to 'tell the truth', or did you hold some information back if you were worried that it might not be correct or valid? In these sorts of conversations, power can be used authoritatively by one member (or group) to elicit specific types of knowledge from another person. While this more top-down approach is perfectly valid, such transactional approaches to interview practice are rarely much fun for the interview participant, and the stress involved can affect the type and quality of knowledge given during the encounter, and occasionally narrow the scope for longer, fuller responses from participants.

Let us flip the scenario and reflect on a conversation where you have been more at ease, say, in a conversation with your best friend, your partner or a work/school colleague you feel equal to. How did the dynamics of that conversation differ from the former? Was there greater balance between who speaks, and the tone and tempo which the conversation took? Did you feel relaxed physically and emotionally when talking? Were you encouraged to speak frankly without fear of repercussions? The chances are you probably experienced that conversation very differently, and contrary to the previous transactional approach, in these types of conversations, power is distributed more evenly to allow knowledge to emerge slowly and in greater detail. In this scenario, you might provide your opinion when posed with a question or challenge someone else's in an inclusive and supportive way. Balanced or equitable conversations are therefore much more effective at understanding people's opinions, motivations and values than an interrogation (see Box 4.2).

Box 4.2   Determining the influence of power in interview design

When designing your interviews, you might want to consider the following:

Who am I interviewing? Your interview participants might be deemed to be powerful in that they have status within society or in an organisation, or they may be vulnerable if, say, they come from a minority group. Think carefully about how your position as a researcher relates to this.

Where am I interviewing them? Check that the interview setting is a mutually safe and comfortable space. Participants might not want to open up about personal matters if being interviewed in a busy public space (see Chapter 6).

What am I interviewing them about? Make sure you can clearly and simply explain your research to your participants and give them some indication of the types of themes you want to discuss. This will help relax your participants and prepare them should they not want to discuss a particular theme with you.

How are my questions likely to be interpreted? Think about your own positionality but also the positionalities of your participants as well. Are you pitching themes or questions correctly to your interview participants so as to have a conversation that is relatable and not patronising?

Could my line of questioning cause offence? Linked to the previous question, think carefully about both the framing of questions, in terms of language, style and content, as well as the delivery. Put yourself in the shoes of your participants and think whether you would like to be asked these questions. If you are concerned that a topic is too sensitive, find ways of gently building up the line of enquiry, and if in doubt, involve your participants in devising the questions themselves.

What happens if something goes wrong? Problems are more common than you might think, and many researchers experience issues when collecting data. If something goes wrong (e.g., if your questions yield only very short responses or a participant wishes to terminate an interview), pause your interviews and think about why the problem occurred and what you might be able to do to improve things for subsequent interviews.

Power imbalances

So far, I have outlined the role of power in research. Yet power can be challenging to negotiate in the professional setting of an interview encounter. While it may sometimes appear that the researcher, by dint of having arranged the interview, would wield the power, there are many academic accounts of researchers who have been intimidated or belittled by participants or even having the focus of the interview intentionally shifted by an interview participant (see research by Vähäsantanen and Saarinen (2013) and Li (2022) for excellent examples of power dynamics between interviewers and elites in Finland and China, respectively). This can, inevitably, have significant effects on the type and quality of data that are produced, as well as the perceived quality of the interview itself, with some accounts discussing such instances as 'failed interviews' (Jacobsson and Åkerström, 2013).

Rice (2010: 70) has written on what he calls "the particularities of power" in interview research. Through a research project that involved interviewing key stakeholders in a retail regeneration development, Rice experienced a range of asymmetric power relations when preparing and conducting the research. These included issues such as difficulties in accessing participants and locations, complications with developing reciprocal relationships with interview participants and problems when trying to retain a sense of control over the themes that needed to be covered. What Rice's experiences can tell us about power is that power is never generalised but is instead produced in particular ways and for particular means. Hence, if left unacknowledged, those who wield power (be they interviewers or interview participants) may have the ability to control what is included and excluded from the research, what knowledge is produced and how this is disseminated.

Box 4.3   Activity – Recognising power asymmetries in interviewing

Choosing a topic that interests you, I want you to think about what it would be like to interview the following interview participants:

- A senior businessperson
- A school child
- A retired person
- A friend or relative

In each scenario, consider the role of power in the interview encounter. What would change in each scenario, and why? Think about this in terms of your behaviour, the way you might ask questions or how you respond to answers. Also, reflect on what types of power (im)balances might occur in each encounter and whether these might affect the quality of the interview data.

To extend this, Mikecz (2012) discusses opportunities to 'rebalance' perceived power imbalances in interview encounters. Using research conducted in Estonia with political and economic elites, Mikecz argues that positionality and power are dynamic qualities in that they shift over the course of the research process and can be proactively influenced and managed by the researcher. In the context of interviewing elite participants, Mikecz encourages interviewers to consider intensive pre-interview preparation to help ensure access, trust and rapport are gained with participants and that the interviewer and interview participants are clear on the application and intention of any interview encounter before it goes ahead. While this might not be possible – or necessary – with members of the public, at least have some background information about your participants (i.e., do your homework!) and establish preferences on the means of communication (i.e., avoiding presuming that a standard interview approach is acceptable to all).

INSIDER/OUTSIDER POSITIONALITIES

So far, I have encouraged you to consider the role of positionality in guiding the research encounter and the power dynamics that may influence how knowledge is produced. In this section, I put these roles into a more practical setting in order to think about how our positions as researchers might situate us as either insiders (i.e., among interview participants who are like us) or outsiders (i.e., among those different from us). Both of these scenarios are methodologically sound, and I will explain how they operate and what to consider when conducting interviews as either an insider or outsider (see Miller and Glassner, 2016; Mullings, 1999). To finish, I will show you an example of how these roles can occasionally blur, and that this might produce different types of knowledge from your interviews.

Insider positionalities – researching 'from within'

It is common for a lot of interview research to be conducted by researchers deemed to be 'within', or close to, the research phenomena itself. We may, for example, have chosen a research topic because it is relevant to our own social or cultural background. It might speak to our values or beliefs, or because we think a particular voice, issue or cause has not been researched before. We might also choose a research topic because we understand the issue intimately – perhaps we are a member of a marginalised group and therefore understand the challenges faced by our potential research participants, or we might be experts in a certain activity that warrants academic research into (see Philo et al., 2021).

Words like 'empathy' and 'credibility' are often used in conjunction with discussions of insider status, and indeed, there is considerable value in having intimate knowledge of a research problem that other researchers may not necessarily understand how to access or make sense of. Insider knowledge of a phenomenon might also allow you greater opportunities to identify and reach out to your chosen sample. Moreover, you may have a clearer idea of the types of themes and questions that you will need to talk to your participants about. You might also find that your positionality might make your participants feel more comfortable during the interview encounter itself. Hence, there can be related personal benefits from conducting research as an insider, and you might find things out about your own practices, beliefs and values that you perhaps did not consider important before the research.

Outsider positionalities – researching 'others'

Researching from an outsider perspective is perhaps a more complex position for researchers to adopt. Those with more quantitative research backgrounds are likely to consider an outsider perspective as a more detached and scientific way of researching a phenomenon – even though it is largely impossible to interpret language, experience and perception from an objective standpoint (Holmes, 2020). By not having prior knowledge of the population or phenomenon under investigation, the researcher 'could' be assumed to be less likely to influence the direction of the interviews and the knowledge that derives from this. Conversely, an alternative school of thought would argue that being an outsider means not having the ability to empathise with the subject group, therefore making it impossible to claim to fully understand the experiences of those being interviewed. Hence, while being an insider infers belonging to the same place, community or social/cultural group as your participants, an outsider perspective could potentially present a barrier to accessing and producing knowledge. Yet, social researchers might consider outsider positionalities as an opportunity to gain new knowledge through which to support the experiences of their interview participants (e.g., opening up new debates around a pressing phenomenon or giving a voice to an under-represented group). What is key when considering interviewing participants from an outsider status perspective is to be honest and open with your interview participants about the research process throughout the planning, practice and dissemination stages of the study.

'To be, or not to be' an insider?

Dwyer and Buckle (2009) draw upon Shakespeare's famous 'to be, or not to be' quote in order to critically examine how the apparently dichotomous positions of the 'insider' and 'outsider' researcher interrelate with one another. Crucially, they argue that viewing insider and outsider positionalities as oppositional is overly simplistic and that we, as researchers, are much more likely to occupy positions that sit somewhere between the two. We are unlikely, for example, to have precisely the same membership to a group or community as our participants – who are also very likely to have a wide range of differences and similarities in their perceptions and experiences of the phenomenon under investigation. We must therefore recognise the things we may have in common with our research participants alongside the things that set us apart. As Tarrant (2014) argues, our insider/outsider positionalities are complexly entangled with various competing power dimensions that mean we rarely experience or view the world in precisely the same way as each other (see also

---

Box 4.4   Activity – Considering positionality in interview design

In this activity we want to try and understand how different groups of participants might understand, perceive or experience a phenomenon in different ways and how participants' (and our own) various positionalities might influence how we view the world.

Choose a fairly generic interview question – e.g., "tell me about how you use your local high street or mall".

Then consider how participants' different social identities (e.g., gender, age, ethnicity, sexuality) might influence how they respond to this question.

List some of the challenges and opportunities that participants with different social identities might face in relation to your question and think about:

- What differences exist within and between these social identities in relation to your question?
- Why do these differences exist?
- How might you frame the question differently to support these social identities?

Mullings' (1999) work on inter-cultural perceptions, interactions and representations among interview participants in Jamaica). Hence, interview research situations can often be both very familiar but also strangely different for researchers and participants. In a study of political relations in Kirkenes, North Norway, Medby (2023) argues that considering positionality reflexively allows interviewers to work 'with' interview participants to produce more dialectic, co-produced knowledge (i.e., knowledge that is formed through discussion) rather than adopting the position of the distant researcher that documents 'their' (i.e., the participants') experiences and perceptions. It is thus important to bear in mind that it may well be that your positionality becomes an important (and valuable) part of your analysis, therefore, do not discount its importance.

REFLECTING ON THE PROCESS – USING RESEARCH DIARIES EFFECTIVELY

In building upon these discussions of positionality and power, I hope that this chapter encourages you not to march straight into collecting data and instead, to build time and space for reflection into the research design process to help understand (a) the progress of the data collection and anything that might have changed during this time, and (b) if, and how, this affects the quality of the data being produced. You will spend a lot of time thinking about these points, probably when you would rather be doing other things, but in order to make these reflections effective and meaningful, you will want to consider ways of documenting your thoughts. A very simple strategy to adopt is to keep a research diary to document the progress of the research process (see Box 4.5). This may immediately sound onerous or 'another thing to think about', but noting a few thoughts after each interview encounter and returning to these notes before conducting the next can help tighten the research process and identify any new or interesting points that you may have overlooked in the design stage.

A note from the field

I first used a research diary during my PhD. My methodology involved conducting walking interviews, whereby I accompanied

Box 4.5   The benefits of using research diaries

A research diary can have four key benefits.

1. Diaries can provide opportunities for pause and reflection that are vital in evaluating interview techniques and the quality of data. It is unlikely that you will have the opportunity to conduct your interviews in isolation from any other activities. Thus, a diary can document your thought processes, achievements and concerns after each encounter.

2. Diaries can help confront and rectify issues that may arise from the interviews themselves, such as hostile or unexpected responses to certain questions. This is not about searching for flaws in your interview technique. Reflecting on a perceived problem might have a practical outcome in refining the interview application, but it may also help you understand more about how knowledge about your phenomenon is produced.

3. Diaries can help explore new ideas or themes that emerge from the research. It is easy to get caught up in the interview process and miss or forget about unexpected findings until you start to transcribe. Sometimes new information might arise from an interview that makes you want to rework some of your lines of questioning or even revisit your research objectives.

4. Diaries allow you to reflect on your own experiences of research, particularly your role as an interviewer. You might be thinking that you only intend to conduct interviews for the purpose of an academic study, yet many occupations contain research dimensions, meaning the skills you gain from this research may well lend themselves to your future career.

my participants on an urban walk and discussed their perceptions and experiences of the local environment (see Holton and Riley, 2014; Riley and Holton, 2016). During this process, my participants also took photographs of the spaces we walked through, and I captured the route of the walk using a Global Positioning System (GPS) device. As this involved a number of methods, I knew I would find it potentially difficult to remember how each encounter went, so I elected to keep a research diary to record some of the more pertinent points of each interview not picked up through the audio recording. This worked in two ways. First, it allowed me to start to pick up any linkages and discrepancies between each participant (for example, I noted in my diary that first-year students tended to choose routes very close to the university campus, compared to their peers in later years who chose much more varied routes around the city). Moreover, it inspired critical and reflective thinking over the success of employing the walking interview method itself. (I found that some encounters were adversely affected by the weather, meaning I often had to be mindful of how the environment might affect the quality of my recordings.)

My research diary therefore became an unexpected, but powerful, management tool in that it helped me evaluate the quality and direction of my project (Valentine, 2005; Silverman, 2015; 2017). I completed it using a word processor package after each encounter. I documented key features of the interview, such as poor weather, new lines of enquiry or ways that the participants engaged with the environment (see Figure 4.1), ensuring they were addressed and overcome in subsequent interviews. My diary was also not just written. I included annotated maps of each route which acted as a visual *aide-mémoire* when reflecting back on each of the encounters. These maps contained noteworthy components of each of the routes taken and were annotated with the times of particularly significant

Date	15/11/2011	Duration	1 hour 48 mins
Name	Liam	Age	21
Destinations	Library, Seafront and Shopping precinct		
Weather conditions	Warm, sunny and dry		
Engagement with interview subject	Fully engaged with subject matter, though likely to be because geography is his degree subject. Gave well rounded answers and talked comfortably about both his social life and home life. Spoke very candidly about how he felt his social position sat within Portsmouth. Also spoke reflectively about his transition into and through university.		
Engagement with environment	Spoke about his connection with spaces both when prompted and voluntarily. Gave plenty of examples of how he related to particular spaces, most commonly the shopping precinct which was a source of much of the social activities data. Interestingly the library spaces were viewed negatively. It was almost as if Liam felt out of place in this area, particularly in the front courtyard in which he dismissed the space as being "a smoking area, full of people smoking and leaving their bikes, so no real attachment for me whatsoever". This is supported by him explaining that this space made him feel alien.		
Predominant features of interview	He spoke confidently about how his social life had adapted over the course of his degree. Being in a walking interview situation allowed me to use the space to prompt deeper analysis of his experiences, particularly if they began to contradict an earlier statement.		

Figure 4.1 An example of a research diary entry

moments of the encounters. These times corresponded with the audio files, allowing key sections of the interviews to be revisited during the analysis.

During the research, I regularly returned to my diary before new interview encounters, reading and re-reading entries to get a feel for how the interviews fitted together as a whole. Reading back through this diary more than a decade on from conducting the research I can see that using it meant I got to know my project extremely well. I could see myself grow in confidence as a researcher and I could see instances where I had to make changes or respond to problems and how I strengthened my methods in doing so.

My diary therefore became a valuable reflexive tool which encouraged me to be critical (in a productive way) about the success of each interview and continually improve my practice. This prevented me from becoming complacent and potentially missing valuable data. A research diary is therefore more than simply a place to deposit reflections and ideas. As Engin (2011: 297) argues, "[A] research diary can be viewed as a scaffolding tool in the construction of both research knowledge and identity as a researcher". I determined then that the implementation of a research diary greatly improved the quality of my data collection because it developed a more reflexive understanding of how my actions as a researcher shaped the findings of my research.

SUMMARY POINTS

- The design, implementation and analysis of interviews are all influenced by the position of both the interviewer and interview participant. The backgrounds and motivations of an interviewer and interview participant relate to one another in specific and unique ways when producing knowledge.
- Positionality should not necessarily infer bias but should instead be acknowledged as an important marker for how our individual worldviews are produced and how these might shape our encounters with others.
- Power plays an important role in interview encounters and can influence the production of knowledge, both positively and negatively. Balanced or equitable conversations are therefore much more effective at understanding people's opinions, motivations and values than an interrogation.

- Insider positionalities are sometimes viewed as empathetic, as the researcher may be considered more able to understand the types of themes and questions that matter to their participants.
- Outsider positionalities can be considered more objective or scientific in that the researcher is more detached from the phenomenon.
- Research diaries can be an effective method for documenting thoughts and reflecting on the interview process over the duration of a research project.

WHAT TO DO NEXT?

Consider the following:

- Familiarise yourself with the meanings of power and positionality.
- Think carefully about who you are, what constitutes your identity (or identities) and how your social and cultural position might affect an interview encounter.
- Consider the position of your participants. Anticipate whether there might be power imbalances, cultural or social differences or any potential sensitivities associated with the interview topic. If you are in doubt, talk to someone about it or even raise it with your participants beforehand.
- Linked to the previous point, consider whether the research you are doing positions you as an insider, outsider or a combination of both. This, again, can help you understand if and how you might influence the direction of the interview and the knowledge that is produced.
- Set up a research diary as early on in the project as you can. This can act as both a repository for notes and ideas and as an opportunity to reflect on your own practice, and the direction and quality of the research.

Suggested further reading

Dwyer, S. C., and Buckle, J. L. (2009). The space between: On being an insider-outsider in qualitative research. *International Journal of Qualitative Methods*, 8(1), 54–63.

These authors grapple with the question of whether researchers should be part of the populations in which they study. They conclude that research is most successful when both insider and outsider positions are considered interlinked.

England, K. V. (1994). Getting personal: Reflexivity, positionality, and feminist research. *The Professional Geographer*, *46*(1), 80–89.

England examines the complex relational identities associated with the researcher and the participant and how these identities influence the types of knowledge that interviews produce.

Jacobsson, K., and Åkerström, M. (2013). Interviewees with an agenda: Learning from a 'failed' interview. *Qualitative Research*, *13*(6), 717–734.

Drawing on the experience of one encounter from a set of interviews based on neighbourliness in Sweden, these authors consider the context of interview participants who might use their power to influence the direction and outcome of an interview encounter and whether this constitutes a 'failed interview'.

Mikecz, R. (2012). Interviewing elites: Addressing methodological issues. *Qualitative Inquiry*, *18*(6), 482–493.

Mikecz provides some valuable advice on how to overcome the challenges of potential power imbalances when interviewing participants from elite backgrounds.

REFERENCES

Adriansen, H. K., and Madsen, L. M. (2014). Using student interviews for becoming a reflective geographer. *Journal of Geography in Higher Education* *38*(4), 595–605.

Cochrane, A. (1998). Illusions of power: Interviewing local elites. *Environment and Planning A*, *30*(12), 2121–2132.

Dwyer, S. C., and Buckle, J. L. (2009). The space between: On being an insider-outsider in qualitative research. *International Journal of Qualitative Methods*, *8*(1), 54–63.

Engin, M. (2011). Research diary: A tool for scaffolding. *International Journal for Qualitative Methods*, *10*(3), 296–306.

England, K. V. (1994). Getting personal: Reflexivity, positionality, and feminist research. *The Professional Geographer*, *46*(1), 80–89.

Holmes, A. G. D. (2020). Researcher positionality: A consideration of its influence and place in qualitative research – A new researcher guide. *Shanlax International Journal of Education*, *8*(4), 1–10.

Holton, M., and Riley, M. (2014). Talking on the move: Place-based interviewing with undergraduate students. *Area*, *46*(1), 59–65.

Jacobsson, K., and Åkerström, M. (2013). Interview participants with an agenda: Learning from a 'failed' interview. *Qualitative Research*, *13*(6), 717–734.

Kezar, A. (2002). Reconstructing static images of leadership: An application of positionality theory. *Journal of Leadership Studies 8*(3): 94–109.

Li, L. (2022). How to tackle variations in elite interviews: Access, strategies, and power dynamics. *Qualitative Research*, *22*(6), 846–861.

Medby, I. A. (2023). Words beyond 'data': Managing small talk and positionality in North Norway. *Area*, *55*(2), 221–226.

Mikecz, R. (2012). Interviewing elites: Addressing methodological issues. *Qualitative inquiry*, *18*(6), 482–493.

Miller, J., and Glassner, B. (2016). The 'Inside' and the 'outside': Finding realities in interviews. In Silverman, D. (Ed) *Qualitative research* (4th ed., pp. 51–66). London: SAGE.

Mullings, B. (1999) Insider or outsider, both or neither: Some dilemmas of interviewing in a cross-cultural setting. *Geoforum*, *30*(4), 337–350.

Philo, C., Boyle, L., and Lucherini, M. (2021). Researching 'Our' People and Researching 'Other' People. In von Benzon, N., Holton, M., Wilkinson, C., and Wilkinson, S. (Eds) *Creative methods for human geographers* (pp. 35–48). London: SAGE.

Pini, B. (2005). Interviewing men: Gender and the collection and interpretation of qualitative data. *Journal of Sociology*, *41*(2), 201–216.

Rice, G. (2010). Reflections on interviewing elites. *Area*, *42*(1), 70–75.

Riley, M., and Holton, M. (2016). *Place-based interviewing: Creating and conducting walking interviews*. London: SAGE.

Secules, S., McCall, C., Mejia, J. A., Beebe, C., Masters, A. S. L., Sánchez-Peña, M., and Svyantek, M. (2021). Positionality practices and dimensions of impact on equity research: A collaborative inquiry and call to the community. *Journal of Engineering Education*, *110*(1), 19–43.

Shurmer-Smith, P. (Ed.). (2002). *Doing cultural geography*. London: SAGE.

Silverman, D. (2015). *Interpreting Qualitative Data* (5th ed). London: SAGE.

Tarrant, A. (2014). Negotiating multiple positionalities in the interview setting: Researching across gender and generational boundaries. *Professional Geographer*, *66*(3), 493–500.

Vähäsantanen, K., and Saarinen, J. (2013). The power dance in the research interview: Manifesting power and powerlessness. *Qualitative Research*, *13*(5), 493–510.

Valentine, G. (2005). Tell me about… Using interviews as a research methodology. In, Flowerdew, R., and Martin, D. (Eds) *Methods in human geography: A guide for students doing a research project*. Harlow: Pearson Education Ltd.

ETHICS AND ETHICAL ISSUES

Chapter objectives

This chapter attends to the importance of recognising and ensuring confidentiality and anonymity when using qualitative interviewing approaches. By reading the chapter, you should

- understand how and why research ethics are important to all interview research;
- develop knowledge of informed consent and how to obtain this from interview participants;
- identify a range of common ethical dilemmas that interviewers might face when designing, practicing and analysing interviews;
- appreciate the importance of effective communication with interview participants; and
- acknowledge the responsibility of the researcher to prepare, secure and store interview data.

INTRODUCTION

So far, I have introduced you to the practical implications associated with designing interview-based research. This has involved establishing a research problem to investigate, identifying gaps in the academic literature, and choosing and preparing an appropriate methodology to investigate your phenomena. But how do we know whether the interview approach we are proposing is correct? What if the questions we ask offend our participants? Do we know whether the location we intend to research in is safe for both us and our interview participants? Have we ensured that our participants'

DOI: 10.4324/9781003292784-6

responses are anonymous and confidential? These are all important questions for researchers to be asking, and all academic research is bound by an obligation to be safe and ethically responsible. Hence, research should never proceed from gut instinct alone. Instead, as researchers, we must devise and adhere to a set of guidelines that ensure the quality of the methodological design carries forward into how we conduct ourselves during the data collection and manage the data provided by our participants throughout the duration of the project and beyond.

This chapter attends to the importance of recognising and ensuring confidentiality and anonymity when using qualitative interviewing approaches. This includes a critical discussion of whose ethical and legal rights are being considered in the interview encounter. I will introduce the reader to why attending to ethics and risk are both important in interview-led research and, subsequently, how ethics and risk assessment can help support both participants and researchers throughout the research process – including preparing, collecting, transcribing and writing up interviews.

I begin by outlining what we mean by research ethics and the ethical frameworks that underpin all research. Next, I focus specifically on the notion of informed consent, what this entails and the steps involved in obtaining consent from participants and sticking to it. Third, I unpack some of the key ethical issues that exist within interview-based research, including harm, deception, privacy, and anonymity and confidentiality. Finally, I outline the importance of effective and clear data management to ensure quality, security and confidentiality.

WHAT ARE RESEARCH ETHICS?

So, what are research ethics? When we think about ethics, we are often drawn to notions of being ethically minded in order to take responsibility for our actions. But what does 'ethical responsibility' actually mean in the context of research and legal frameworks? From a UK perspective, the Wellcome Trust (2014: 1) defines research ethics as "[…] the moral principles that govern how researchers should carry out their work". This notion of principles hints at ethics as a framework for scaffolding responsible actions into all stages of the research. Clark et al. (2019: 122) go on to posit five key ethical concerns for researchers to consider when designing a project:

- Informed consent
- Confidentiality
- Anonymity
- Avoidance of harm
- Privacy

While these each have different requirements, they ostensibly link together into an overarching responsibility to clearly articulate the processes involved in the proposed research, what is required from participants who take part in the research, how the outcomes of the research will be reported and, most importantly, the steps that need to be taken to ensure participants are protected from any potential harm if taking part in the research.

I documented in Chapter 2 how qualitative researchers need to ensure rigour and responsibility in their research design, in much the same way that quantitative researchers do through scientific and replicable experiments. Ethics is one of the essential tools that ensure that qualitative research is conducted responsibly and with integrity, and therefore ethical protocols must be built into all social research applications at all stages of the research. Yet, it is not enough to just simply say that you will act responsibly when designing, conducting and writing up your interview research. Education institutions, research councils, charities and other organisations, all have ethics committees to whom formal ethical approval documents must be submitted, reviewed and signed off before research can be conducted. Research ethics are therefore highly regulated, being couched in formal legal frameworks designed to make researchers aware of their responsibilities towards the participants they work with and the settings in which research takes place. Wiles (2013: 1) calls this 'ethical literacy', arguing that obtaining ethical approval should not be considered a hurdle to overcome before commencing a project but as a way of thinking critically about the process of research as it unfolds and how ethics are aligned and adapted to this.

At this point, you may now be worrying about ethics and harm in your own interview research. There are, of course, many horror stories concerning unethical research – a classic example being Stanley Milgram's shock experiments in the 1960s in which participants were encouraged to administer 'fake' electric shocks to each other to test people's obedience to authority (see Clark et al., 2019). This is an extreme example, and the majority of ethical issues that researchers face are fairly mundane and everyday by comparison. They are no less important, though, meaning understanding the context of ethics and what constitutes an ethical issue is a vital step in ensuring all aspects associated with your interviews are conducted properly and safely.

Informed consent

I will start this discussion of ethics by examining informed consent. It is no accident that informed consent is placed first on many lists of ethical considerations. While there are contestations surrounding consent and its implications for affecting 'natural' research (Israel, 2015), the majority of researchers agree that participants should be given the right to choose whether to participate or not in a study (Robson, 2002). Hence, within interview-based research, informed consent operates as the primary mechanism used to ensure potential interview participants understand the steps involved in their participation, what they are obliged to do if they choose to take part in an interview and what you intend to do with the information they entrust you with during the process. There are important steps involved in obtaining informed consent from interview participants and gatekeepers, including how to collect and store information and how to support potentially vulnerable participants.

Informed consent is primarily based upon the notions of transparency and trust. If you are clear about what your research entails, then your participants should also feel secure in the knowledge that you are working responsibly. Informed consent is typically obtained in writing, using a form that briefly introduces the research and the methods involved and then invites participants to confirm that they are happy to proceed with their involvement (some interviewers might do this verbally by audio recording the participant's consent at the beginning of the interview). This often takes the form of a list of points to tick against that is signed and dated by the interview participant. Some institutions may well have templates to use.[1] As Clark et al. (2019: 127) argue, written consent presents a "clear record of agreement" between the researcher and the participant. However, written consent does contain personal information, such as names and signatures, so it is important to ensure such documentation is kept confidential and separate from any interview transcripts or other data.

When preparing an informed consent document, it is important not to make it too formal. Yes, the information will need to be clear, but adopting a friendly tone will help you come across as approachable, particularly if a potential interview participant has questions or wants to seek clarity. What to include in a consent form is also important to consider. If you are using a template, then that can guide you, but you will no doubt need to tailor the form to the specific requirements of your project. Remember that ethics is concerned with how data are collected, recorded, stored and disseminated, meaning, to obtain a participant's full consent, you will need to outline all of the steps involved in your interview procedure so as to guide your participants through their involvement (e.g., include information about the methods of recording, storing and disseminating your data and how this will be achieved). You will also need to ensure that your participants understand what they are consenting to when they complete the form rather than assuming they have fully understood it (see Box 5.1).

Box 5.1   Tips for agreeing to consent

These points are adapted from Robson (2002) and can operate as a type of checklist for preparing and utilising consent forms.

- At the start of the interview, seek clarification from your participant that they understand the project and their involvement.
- Once the interview is completed, check again that they are okay for you to proceed with using their data in your project. This may only be clear to them once they have been through the process.
- If you are using an interview approach that requires participants to be involved multiple times (e.g., interviewing over intervals of time), make sure that consent is obtained at each interval.
- Try getting creative. Sometimes the written word can be confusing, so perhaps consider visual methods like photographs, graphics or video to help illustrate the consent procedure.
- Avoid overwhelming participants with lots of information. Keep consent forms sharp and to the point. Piloting these beforehand can help ensure all information is understandable.
- Do not bully your participant into taking part in an interview. Some participants may need time to absorb the information or want to discuss it with others. They might also not wish to take part. Allow participants space to do this without fear of embarrassment, coercion or comeback.

Finally, it is important not to assume that all participants will be able to provide their consent to take part in an interview. The capacity to consent is crucial for conducting research ethically and is bound by legal frameworks, such as the 2005 Mental Capacity Act in the United Kingdom, which ensures vulnerable groups are protected from harm in research contexts. Vulnerability is quite a broad term; however, in research, this often refers to children and young people under the age of 18, people with intellectual disabilities, and those with certain forms of physical disabilities and/or mental illness (Wiles, 2013). In these contexts, there could be issues surrounding the capacity for participants to fully understand their involvement in research and the consequences of taking part, meaning others may need to be involved in the process of obtaining consent and ensuring the participants are safe and secure throughout the duration of their involvement (Clark et al., 2019). This is termed 'proxy consent', and obtaining consent via these means might involve consulting parents, guardians, carers and/or institutions (e.g., schools, colleges, healthcare providers). The steps involved in Box 5.1 apply in exactly the same way here, except it is likely you will be required to supply an additional informed consent form to whoever is providing the proxy consent. Moreover, it is important to ensure that you have full clearance to access vulnerable groups for your research. In the United Kingdom, for example, researchers must go through a Disclosure and Barring Service (DBS) check in order to work with vulnerable groups – particularly children.[2] This all takes additional time, so remember to plan ahead when designing your research methods to ensure you have sufficient time to complete all of the steps before conducting your interviews.

IDENTIFYING KEY ETHICAL ISSUES IN INTERVIEW-LED RESEARCH

In the previous section, I outlined the methods through which researchers can obtain consent from participants to conduct research in transparent and trustworthy ways. While informed consent is vitally important, it would be wrong to assume that ethics ends at the point of obtaining consent. Ethics are built upon a framework of four other core principles of protection associated with harm, deception, privacy, and anonymity and confidentiality. Collectively, these work together to ensure research is practiced within institutional, administrative, legal and moral guidelines and that participants are aware that any potential ethical incursions have been considered and mitigated prior to their involvement.

Harm

One of the primary ethical considerations involves identifying and mitigating risk within research settings. This comprises thinking ahead to anticipate potential risks but also responding to issues if risks arise (Wiles, 2013). Having a clear plan will, therefore, ensure you are able to identify and dynamically respond to risks throughout the research process. Many institutions – schools, colleges, universities and workplaces – will have a formal way of anticipating, measuring and recording risk, usually involving a risk assessment form. Risk assessment, like ethical approval, fits within a framework of legal, institutional and administrative guidelines and must be considered carefully when designing and conducting all forms of research.

Risk assessment can be defined as a threefold process – (1) as a system of *identifying* potential hazards, (2) through the *evaluation* of the risks associated with the hazard and (3) by *implementing* control measures to minimise the risks identified. In this context, a 'hazard' is anything that has the potential to cause harm (e.g., a hazard would be being run over by a car, rather than the car itself), whereas a 'risk' is the likelihood that someone will be harmed by the hazard (e.g., conducting your interviews while walking across a busy road) (Lofstedt, 2011; Scheer et al., 2014). There are a variety of techniques for assessing risk. The UK Research and Innovation (UKRI) body presents a five-step risk management process that encourages researchers to assess, review and actively manage risk throughout the duration of a project. These include (1) identifying and recording risks at the initial stage of the project design, (2) assessing and analysing the impact of such risks and prioritising the severity of threats, (3) planning actions to mitigate risks, (4) outlining strategies to implement such plans and (5) clarifying approaches to monitoring and assessing the relative outcomes of these strategies (Kingston and Howard, 2017).

In interview contexts, physical risks are fairly easy to determine and mitigate and are likely to refer to the spaces in which interviews take place. These might include environmental hazards, such as terrain or moving vehicles, machinery or objects and climatic hazards, like exposure to heat or cold, or weather events. In addition to these, interview-based research may contain psychological risks, such as handling sensitive topics, framing difficult questions or uncovering traumatic events. In these contexts, our intention is not to harm our participants, yet in asking them to recount events or respond to pressing issues, we may unintentionally generate fear, stress, anxiety or emotional pain for interview participants. Psychological risks are therefore very difficult to pinpoint and may not reveal themselves prior to the interview encounter itself.

Careful planning is therefore a key driver to mitigating psychological harm. Making it clear to participants what themes, topics or questions you intend to cover can help interview participants mentally prepare to respond, particularly if they are aware of any sensitive topics you intend to cover. Yet, you will also need to consider strategies for handling distress during the interview process itself (see Chapter 6). Informing participants of the support mechanisms you have put in place if they were to become distressed before you start to interview them is often a positive, empathetic step. Stopping the interview (i.e., turning off or pausing the voice recorder) and allowing participants the opportunity to step away from the research setting may help provide participants the space to regroup and return to the interview. Some participants may feel it too much to continue though, and at this point, the interview should be terminated and the participant given the opportunity to leave. Providing contact details for relevant, local support organisations is also key here. We are unlikely to have formal training to support those experiencing psychological distress, and intervening may cause more harm than good. It is therefore important to signpost participants to those with professional expertise before any issue arises.

Finally, attention to harm should not only be confined to your interview participants. Your own safety is also paramount when using interviews in research. All risk assessments will require you to consider 'safety in the field', be it in the interview location itself or the methods involved in getting to and from these locations. Protocols for 'checking in' and 'checking out' of an interview encounter should be designed into your risk assessment, as must the rationale for choosing a research location (Dickson-Swift et al., 2008). Consider, for example, the legitimacy of researching in a participant's home against interviewing them in a public space. You might have developed a strong rationale for visiting participants to interview them at home, but what additional measures would you need to put in place to ensure your own safety, as well as the safety of your participants (see Chapter 6)?

Deception

It goes without saying that credible research should never deceive or mislead participants. A well-conceived consent form should outline the intention of the interview and all the steps involved up to the point of completion. Deviating from this, or altering the research intentions, would contravene the consent that was obtained and may inflict harm on those involved – including your participants, you as a researcher and any institutions or organisations you are affiliated with. Deceptive interview practices can conjure images of tabloid news reporters looking for juicy scoops, disguising themselves to hide their identities and manipulating the purpose of the interview they are conducting to uncover uncomfortable personal information. This type of intentional deception would be considered very unethical in academic research, and the consequences for conducting research deceptively would be grave (Sieber, 1992). Yet, some research methods may require you to waive written consent. If adopting ethnographic or participant observation approaches, you may feel compelled to preserve, as much as possible, a 'natural' research environment that gaining consent might disturb. Or it could simply not be feasible to obtain consent if observing large volumes of people. In these situations, you might provide an 'opt-out' procedure, whereby everyone in the research setting is considered to be involved in the research, but individuals have the opportunity and right to be exempted from being included (i.e., their practices will not be observed, or their voices recorded). Interview research, however, requires neither of these things – you are likely to be known to your participants if you have recruited them, and the composition of an interview encounter means it is a dialogue between two people, making consent straightforward to obtain.

Privacy

At face value, privacy seems like one of the easiest ethical dilemmas to control in interview research. After all, we have the right to privacy in our everyday lives, so it stands to reason that this should extend to research settings too. Our interview participants are therefore able to exercise the right to decide what personal information they disclose or withhold in an interview setting. This may sound obvious, but bear in mind the power dynamics at play in interviews that I outlined in Chapter 4 and the ways in which the perceived authoritative power of the interviewer might lead participants to feel compelled to divulge information they may not necessarily feel comfortable talking about. This is unlikely to be an intentional act on behalf of the interviewer, but it is important to ensure participants feel at ease enough to trust the researcher with their information, as well as having the confidence to decide for themselves what to offer (Sieber, 1992). A statement on privacy can be written into your consent form and can detail what privacy means in the context of your research, how you intend to handle the disclosure of private information and the freedoms you provide your participants in disclosing or withholding personal information. Being clear with participants at this stage can help build the trust necessary for participants to feel more relaxed and open in the interview encounter.

Privacy also extends beyond the interview itself. Privacy is somewhat bound up with anonymity and confidentiality, both of which I will outline next. It is the obligation of the researcher to ensure participants' privacy is maintained throughout the research process (Hammersley and Traianou, 2012). This extends to if and how you discuss other interview participants' experiences during another interview, how you report your findings in your write-up and how you discuss your interview encounters with others, such as with a teacher or advisor, during conversations about your research. Hence, tips like transcribing audio recordings very soon after an encounter can help you get into the swing of referring to participants by their pseudonyms rather than their given names.

Anonymity and confidentiality

As I mentioned earlier, anonymity and confidentiality are somewhat bound up in notions of privacy. I purposefully link these together here so as to indicate how anonymity and confidentiality should be managed in conjunction with one another and provide guidance on how to manage these processes. In defining anonymity and confidentiality in research terms, anonymity refers to the obligation to protect an individual's identity from being revealed intentionally or unintentionally. Confidentiality concerns the security of information – specifically personal details – provided by participants that must not be shared with others without consent. I discuss anonymity in Chapter 9 in the context of transcribing audio recordings, but the process of ensuring anonymity in interview research starts much sooner than that. Your participants have the right to anonymity from the moment contact is made, hence, identifying potential participants to other participants would be considered unethical practice. Likewise, acknowledging participants by name in your dissertation, thesis or report without their explicit consent would contravene your ethical responsibility to protect anonymity.

Anonymity will require changing the names of the participants themselves, providing what is known as a pseudonym or fake name (see Chapter 9). This will also likely extend to other identifying information like towns, villages, schools, workplaces or community groups. Basically, if there is a possibility that your participant could be easily identified through the information they have provided you, then steps will need to be taken to protect their anonymity. In some circumstances, this may not be possible. I read an academic journal article some years ago that documented the migratory experiences of a young British woman as she moved from the United Kingdom to the Middle East. I had a distinct feeling of déjà vu as I read the accounts of this woman's experiences, and I realised I had worked with her in a previous career. While this is a one-in-a-million chance, I recount this story to reassure you that you cannot completely guarantee anonymity, but you can take the above steps to protect it as much as possible.

Confidentiality is much easier to control, and as researchers, we all take responsibility for ensuring the data produced through our interviews are handled securely and confidentially. This means ensuring participants' information is not shared with others, as well as safeguarding how and where their information is stored. This might take the form of password-protected files stored on computers or cloud servers or physical forms of data storage, such as encrypted hard drives or locked filing cabinets for paper copies of documents. This may all sound a bit like something from a spy movie, but confidentiality is bound within legal frameworks of data protection, such as the Data Protection Act 2018, which is the United Kingdom's implementation of the European Union's General Data Protection Regulation (GDPR). All research must comply with these rulings, so it is important to factor in careful planning for data storage and management when preparing your research plan (for discussion, see the "Data Management Plans" section).

Crucially, there are some subtle variances in what information should be considered confidential and what should not. Seidman (2006) argues against stating promises of 'complete' confidentiality in interview consent forms, as this implies that all of the data within the interview itself could be inferred as confidential, therefore rendering the data unpublishable. Instead, he suggests tailoring confidentiality around the sharing of identifying information (e.g., names, signatures), as well as the handling and storing of audio recordings, transcripts, consent forms and so forth. This helps decouple identifying information from the interview data itself, making it available for analysis and dissemination, all in the context of ethical approval.

Moreover, data sharing is becoming increasingly mainstream, and some research councils, organisations and academic journals request that data files be made available to others through centralised services such as the UK Data Service.[3] Such forms of data reuse involve datasets being archived for use by other researchers. This normally involves fully anonymising interview transcripts and providing clear guidance on how this has been achieved, including parameters for what can and cannot be included in the reuse files. I mention this here, as you would need to inform your participants that data reuse is part of your research process – usually within the consent form – if you intend to make data available.

DATA MANAGEMENT PLANS

I take this opportunity at the end of this chapter to touch on data management. This has been a theme that has run throughout the chapter, and my reason for including this section here is to consider how data can, and increasingly is required to be, managed formally throughout the research process (Burnette et al., 2016). Alongside ethical approval and risk assessment, data management plans are now commonplace across most educational and industry research institutions. Research councils, such as the UKRI,[4] Australian Research Council[5] (ARC) and European Research Council[6] (ERC), among others, each require academic researchers to include a data management plan with their research applications. This is beginning to feed through into doctoral PhDs and master's/graduate research (and, more recently, undergraduate degree dissertation research proposals), as well as being adopted by large global organisations like IBM.

While this might appear like an additional hurdle that is being put in the way of 'doing' your interviews, we must be mindful of our responsibility as researchers for the effective preparation, storage and security of our data. Where previously interviews might have been recorded on cassette tapes, transcribed using typewriters or simple word processors, and the paper outputs stored in locked filing cabinets, the technologies we rely on today, while making it much easier to record, transcribe and store data digitally, present a raft of additional problems when it comes to data security and management (see Box 5.2).

Box 5.2   The data management process

Knight (2018) outlines four key steps involved in successful data management that I have adapted for interview-based research:

1. Preparing for collecting data – this involves planning interview questions, the methods through which audio and/or visual material will be captured, data handling 'in the field' and piloting interview methods.
2. Preparing data for analysis – this comprises strategies for securing data storage, methods for data transcription, anonymising and redacting any identifying information, identifying analysis software types and handling consent forms and other sensitive documentation.
3. Preparing for preservation – here we consider the duration of time after the completion of a project that interview data must be kept and the formats it will be preserved in (e.g., audio (mp3, .wav) text (.doc, .txt. pdf) or images (mp4, jpeg, PNG)).
4. Preparation for data sharing – if data resources are to be shared, then document any methods for data sharing: what can and cannot be shared, what permissions need obtaining to share data, where data will be stored, who manages this process, how researchers can access data resources and how these resources are permitted to be used.

There are also practical benefits to producing such a plan. Stating clearly what data you intend to produce and what format(s) these will take helps acknowledge the parameters of the research project. Your data management plan might identify if your project plans are perhaps a little ambitious, as well as the feasibility of transforming interview data into a format that is practical to analyse. If you are using other methods alongside your interviews, you might be able to see if and how these methods link together so as to establish complementary analysis techniques across the datasets. Your plan will also require you to consider whether or not you have a storage facility of sufficient size that can contain your data and how you will access it. These types of statements may sound obvious, but I have read many research proposals that have not thought this through and therefore could become very problematic or time-consuming further down the line if elements of the project need refining.

SUMMARY POINTS

- All research must be designed, conducted and analysed according to a clear ethical framework that respects the rights, dignity and confidentiality of participants while minimising risk and harm.

- All information – particularly information about the study, what participants are expected to do, and what will happen to their data – should be conveyed to participants in a clear format and using understandable terms.

- Participants should be given the opportunity to choose to consent to take part in a study or not.

- Research must be designed to ensure key ethical issues, such as harm, deception, privacy, anonymity and confidentiality, are considered, and strategies to mitigate the negative consequences of these are designed into the project.

- Good interview-based research is built on successful communication between researcher and interview participants at all stages of the research process.

- Effective and clear data management is a key priority for successful research, and it is the responsibility of the researcher to prepare, secure and store interview data.

WHAT TO DO NEXT?

Consider the following:

- Make sure you familiarise yourself with your institution's, or organisation's, ethical policy and process, as well as the methods for identifying, assessing and documenting potential risks.
- Think carefully about the implications for harm, deception, privacy, anonymity and confidentiality in your research. This will be specific to your own project, so you might need to talk it through with a teacher or advisor.
- Design an informed consent form that outlines your project and details the specific requirements of your participants. Provide statements on how data will be produced, stored and used, and make sure participants agree to these terms before continuing.
- Generate a plan for managing your data. Your institution or organisation may have a specific strategy for doing this, but if not, think carefully about how you will manage the data, store it anonymously and retain confidentiality throughout the process.

Suggested further reading

Clark, T., Foster, L. and Bryman, A. (2019). *How to do your social research project or dissertation*. London: Oxford University Press.

Chapter 8 of this book provides a comprehensive guide to managing ethics in social research, with a clear five-step set of principles for managing ethical practice.

Dickson-Swift, V., James, E. L., Kippen, S., and Liamputtong, P. (2008). Risk to researchers in qualitative research on sensitive topics: Issues and strategies. *Qualitative Health Research*, *18*(1), 133–144.

Using examples from Australian research projects, Dickson-Swift et al. switch the focus of risk away from risk to participants to the risks involved with being a researcher and some strategies through which to mitigate risk in research.

Wiles, R. (2013). *What are qualitative research ethics?* London: Bloomsbury.

Wiles sets out a short but thorough approach to considering ethical practice in research. The chapter on informed consent is particularly informative.

NOTES

1 There are many universities and research institutions across the world that have links to informed consent templates that can be used for qualitative research like interviews. The World Health Organisation (2023) is one: https://www.who.int/groups/research-ethics-review-committee/ guidelines-on-submitting-research-proposals-for-ethics-review/templates-for-informed-consent-forms

2 There are many international equivalents, such as the National Police Check (NPC) in Australia.

3 https://ukdataservice.ac.uk/

4 UKRI Data Management Plan: https://www.ukri.org/councils/ stfc/guidance-for-applicants/what-to-include-in-your-proposal/ data-management-plan/

5 ARC Data Management Planning: https://www.arc.gov.au/about-arc/ strategies/research-data-management

6 ERC Open Research Data and Data Management Plans: https://erc. europa.eu/sites/default/files/document/file/ERC_info_document-Open_Research_Data_and_Data_Management_Plans.pdf

REFERENCES

Burnette, M. H., Williams, S. C., and Imker, H. J. (2016). From plan to action: Successful data management plan implementation in a multidisciplinary project. *Journal of eScience librarianship*, *5*(1), 1–12.

Clark, T., Foster, L. and Bryman, A. (2019). *How to do your social research project or dissertation*. London: Oxford University Press.

Dickson-Swift, V., James, E. L., Kippen, S. and Liamputtong, P. (2008). Risk to researchers in qualitative research on sensitive topics: Issues and strategies. *Qualitative Health Research*, *18*(1), 133–144.

Hammersley, M. and Traianou, A. (2012). *Ethics in qualitative research*. London: SAGE.

Israel, M. (2015). *Research ethics and integrity for social scientists: Beyond regulatory compliance* (2nd ed). London: SAGE.

Kingston, P., and Howard, L. (2017). Project risk management quick reference guide. Available on the *UK Research and Innovation* website https:// www.ukri.org/wp-content/uploads/2022/05/STFC-230522-ProjectRisk ManagementQuickReferenceGuideJune2017.pdf. Accessed 9th March, 2023.

Knight, G. (2018). Data management for interview and focus group resources in health. Retrieved from *the London School of Hygiene and Tropical Medicine*. https://researchonline.lshtm.ac.uk/id/eprint/4646631/1/ RDM4InterviewData.pdf. accessed 9th March, 2023.

Lofstedt, R. E. (2011). Risk versus hazard–how to regulate in the 21st century. *European Journal of Risk Regulation, 2*(2), 149–168.

Robson, C. (2002). *Real world research: A resource for social scientists and practitioner-researchers.* Chichester: Wiley-Blackwell.

Scheer, D., Benighaus, C., Benighaus, L., Renn, O., Gold, S., Röder, B., and Böl, G. F. (2014). The distinction between risk and hazard: Understanding and use in stakeholder communication. *Risk Analysis, 34*(7), 1270–1285.

Seidman, I. (2006). *Interviewing as qualitative research: A guide for researchers in education and the social sciences* (3rd ed). New York: Teachers College Press.

Sieber, J. E. (1992). *Planning ethically responsible research.* London: SAGE.

The Wellcome Trust. (2014). Ensuring your research is ethical: A guide for extended project qualification students. *Wellcome Trust.* Available at https://wellcome.org/sites/default/files/wtp057673_0.pdf

Wiles, R. (2013). *What are qualitative research ethics?* London: Bloomsbury.

PART II
DOING INTERVIEWS

'DOING' INTERVIEWS

Chapter objectives

This chapter examines the processes involved in putting interviewing approaches into practice. By reading the chapter, you should

- understand how to manage interviews in practice,
- comprehend a range of practical considerations associated with interview practice,
- identify and develop mitigation strategies for interferences and safety concerns, and
- be aware of potential problems, such as technical issues and participant distress, hostility and emergencies.

INTRODUCTION

This chapter marks the point at which researchers shift from planning interviews to doing them. It would, of course, be inaccurate for me to suggest that the planning stage is finished, and in some instances, you may need to return to your plans, perhaps to adapt aims and objectives, reapply for ethical approval should changes take place in the research design, or revisit your interview guide if tweaks need making to questions. Putting that aside, this chapter is devoted to the application of 'doing' your interviews, and I will take you through some of the practical steps involved in lifting your interview design from the page into research practice. A word of caution though. Unlike quantitative research, which usually has rigid rules as to how processes need to be applied, interviews are extremely

DOI: 10.4324/9781003292784-8

flexible in nature, meaning there is no clear blueprint for the number of interviews you need to conduct, the length of an interview or the way that it is conducted (see Brinkmann and Kvale, 2018). There is likely to be a lot of variation in how interviews are actioned that will be specific to your own project. I will therefore use this chapter to provide guidance and support on what 'could' be done for your research. I would, therefore, suggest making sure that before you get on with doing your interviews, you test your interview design with a set of pilot interviews (see Chapter 3) to make sure the approach you take is appropriate and achievable.

In moving into the doing phase of interview practice, there are a series of practical considerations that you will need to give thought to. Your timeframe for collecting data will be important, particularly in ensuring you have enough time to interview your participants while not burning out with exhaustion in the process. Very practical steps, such as understanding when and how to record an interview, whether to take notes, how to listen and respond to participants and what to do with silences or interruptions, are all important elements that we, perhaps, sometimes presume we are able to naturally do. As you are working with human participants, there may be variance in how some of these steps are actioned. For example, some participants might not want their interviews recorded, while others may request a last-minute change in the interview location. Also, participants' personalities will differ, meaning some participants will be warm and smiley, while others could be very nervous or shy. The key to getting all of this right is clear communication, and this is the theme I want you to keep in mind as you work through this chapter. Asking participants if they are comfortable with the voice recorder being turned on or checking that they understand a question will usually pay off with a smoother-running interview encounter and will often mean that participants will be more likely to check back with you if they are unsure or uncomfortable.

I begin with advice on how to manage interviews in practice, focussing on scheduling, voice recording and note-taking. Next, I outline some of the practical considerations, such as building rapport, listening and dealing with silences and non-verbal cues. Third, I examine the implications for interferences in terms of dealing with noise and other people outside of the research context. Finally, I provide some advice on what to do if things go wrong, focussing here on overcoming technical issues and how to handle distress, hostility and emergencies.

INTERVIEWS IN PRACTICE

In this section, I examine the practical steps involved in putting interviews into practice. The three steps I cover – scheduling, voice recording and note-taking – span the entire 'doing' phase, so it is important to consider them carefully to make sure you get the most out of each encounter.

Recruitment and interview scheduling

Recruitment and interview scheduling is often overlooked in interview guidance. It is commonplace in the literature to carefully detail the planning and sampling stages and then jump straight into 'doing' the interviews themselves. Recruitment and scheduling are key components to the interview process, and developing a strategy to know 'how' to recruit, 'when' interviews are to be arranged and 'where' they are to be held will help manage the interviewing process.

I outlined sampling in Chapter 2, so you should have identified who your target audience is likely to be. The next step will be to recruit potential participants from this sample. I use the word 'potential' intentionally here, as not all of those in your sample are likely to want to take part in the research. Response rates can often be very low for qualitative research, and it can feel quite demoralising if you are unable to reach your participant group. Having a well-crafted and professional recruitment letter or email will usually help improve recruitment, and I have outlined some tips for structuring one in Box 6.1. You can also often append a letter as an attachment in social media or chatroom posts, too, if you are recruiting online.

When scheduling interviews, time is an important aspect to consider. It is important to understand the logistics of your specific interview process to ensure that you are scheduling enough time to complete the interview itself – particularly important if your interview includes activities or lots of themes – and that you have agreed to this with your participants when scheduling the interview. It would be unreasonable to expect a participant to change their

Box 6.1 Designing an invitation letter

Below are some tips on composing an invitation letter for prospective interview participants. This is not an exhaustive template, so consider what is the most valuable and relevant information for your own project and tailor the letter to suit. Aim for around one page of A4 paper.

You are effectively 'cold-calling' potential participants, and they are unlikely to know who you are. To ensure you come across as a credible researcher, structure your invitation as a formal letter or email and use professional language and terminology.

That said, you will need to make your project understandable, so write simply, to avoid complicated words or phrases and explain technical terms.

Start by outlining the title and aim of the research, and include your full name.

Next, introduce the project. State the following:

1. Why you are doing it
2. What you want to find out
3. Why they have been approached to take part
4. What benefits could come from their taking part

Then, describe what is expected of a participant. Outline the following:

1. The method you are using and how they will take part
2. Where the interview will be held, when this is likely to happen and how long it will take
3. What the data will be used for

Include a statement about privacy and anonymity (see Chapter 5), and explain how you will gain their informed consent and how you will store their data.

Explain that the participant's involvement is voluntary and that they have the right to withdraw from the project at any point of the research process (you may want to include a cut-off date here, as it will be impossible to remove data once the research is written up).

It is good practice to include contact details for any complaints or further queries, as well as details of counselling or support services within the vicinity of the research.

Finally, thank the participant for reading the letter and sign off appropriately.

plans if you were to over-run, and you want to avoid missing valuable information if you end up having to rush an interview. There is no fixed time limit for an interview, but many researchers would advocate setting aside between 30 minutes and 1.5 hours for discussion. The general rule of thumb is that too short, and the interview is unlikely to yield results of sufficient depth, too long, and the participants are likely to lose concentration or become bored.

Moreover, try to act quickly when recruiting participants. If a potential participant expresses interest in taking part in an interview, avoid leaving it too long to respond or schedule too far in the future. Your participants may not be as enthusiastic further down the line or may forget to take part. Remember that it is the participant's right not to take part, so avoid pestering or pressurising them into an interview. If they stop responding to your invitations then it is their choice to do so. Finally, think carefully about how long it will take you to conduct all of your interviews. Depending on the volume of interviews you are aiming towards, as well as the time constraints of your project, you will need to set aside sufficient time to 'do' the research. Interviewing is both time-consuming and tiring, so it is unlikely that you will have the time or concentration to conduct lots of interviews in a single day or even packed in over consecutive days. You might consider phasing the process by focusing on a target number of interviews, assessing the results of these and then pressing on with the next phase. This can make the process both manageable and allow for necessary tweaks to the interview questions of themes.

The other aspect of scheduling is location. It is tempting to think that interview locations do not matter, but as I outline later in this chapter, the setting of an interview can influence the success and failure of an interview in many different ways. Many interview guides will advocate establishing a location that is convenient to both the interview participant and interviewer. This might involve a 'neutral' space, like a coffee shop or public library, or a necessarily convenient location, such as a place of work or study. Interviewing in public environments is often perceived as a safer and less threatening prospect for both interview participants and interviewers in that they can usually produce less formal and more natural conversations. Places of work or study may, for example, carry connotations of being on someone else's 'territory', which, again, can affect the direction of the interview itself. I will outline interferences later in the chapter, but crucially, safety must be considered when choosing a location to interview. You should not arrange to interview participants in locations that make you feel uncomfortable or that could put you in danger. Some researchers advocate interviewing in participants' homes, for example, and while this might be a comfortable setting for a participant, there are safety concerns for you as a researcher. So, when conducting an interview in person, always make sure you have a form of identification to prove you are who you say you are. Tell someone where you are going, what you are doing and your anticipated return time and provide them with a mobile telephone number that you can be reached on.

Recording interviews

One of the crucial tools for interview-based research is likely to be a voice recorder. This may seem like an obvious point to raise, but interviewers need to think carefully about how an interview discussion is going to be recorded, whether the technology is appropriate for the activity and how technology, like voice recorders, might be received by interview participants.[1] Capturing an interview using a voice recorder is often taken for granted as best practice in interview research, yet, in doing so, are we perhaps assuming that all of our participants will automatically be okay with this? You will probably recall witnessing media footage of politicians and celebrities who declare 'no comment' when a reporter's microphone is thrust under their noses. A research interview can feel much the same for some participants, particularly if they are nervous about talking or feel unclear about the motivations of the research. Hence, while this section explores the technology itself, I also encourage you to carefully consider the role of the voice recorder in producing (and reproducing) forms of power and positionality in research (see Chapter 4), and I will illustrate how you might alleviate some of the concerns that participants may have in relinquishing control of their own words, as well as worries about privacy and data sharing.

If you were reading this book in the 1980s or 1990s, I would be advising you to conduct your interviews using an analogue hand-held Dictaphone or table-top tape recorder that records the dialogue onto physical cassette tapes that would later be transcribed manually. This technology was often prone to static interference, or the tapes could be susceptible to damage or loss. Technology has, thankfully, improved, and there is now a vast array of digital voice recorders, including hand-held devices, computer programmes and mobile phone applications, that reduce noise interference, are straightforwardly operated and can be easily transferred to other devices and storage locations. Recording an interview now is, therefore, potentially much easier and has greater portability.

Hayes (2021) argues that the status of the voice recorder has become a somewhat ubiquitous and central tool in the interview toolkit, making it almost as crucial as the voice itself in that it provides a conduit from the spoken to the written word. Relying on memory is, of course, impossible, and even taking written notes can miss valuable information – such as an interview participant's precise response – no matter how comprehensively recorded they are. We might, in fact, think it almost impossible (unethical even) to conduct an interview without depicting a participant's words verbatim. A recorded interview not only captures the participant's voice but also your own. It holds valuable information about tone and non-verbal information like laughter, as well as pauses and silences. We therefore take it for granted that our participants will share our enthusiasm for using a recorder. In my own practice, I take time at the beginning of each interview to describe what recording involves, how I will use the recording (as well as how and when I will not), and what will happen to the recording once the interview is finished. In face-to-face interviews, I always let the participants decide on where they want the recorder to be situated, as sometimes automatically placing the recorder on the table in front of a participant can come across as both threatening and distracting.

At the other end of the interview, we also need to think carefully about how and when to conclude the encounter. Basically, when is it appropriate to turn the recorder off and stop collecting data? It may sound fairly obvious that the interview will need to end at some point, and you hopefully see yourself as polite enough to know how to thank someone and part company. My motivation for including this point is to consider the ethics and morals surrounding what concluding an interview actually entails. I was horrified to read a Tweet some years ago that advocated covertly leaving the voice recorder running after the interview had finished as this was often the point that interview participants would offer their most interesting opinions. Never do this, as it compromises the relationship you have built with your participants and would constitute an ethical breach. Once you have informed your participant that the interview has finished, then turn off the recorder and tell your participant that you have done this. If the conversation continues, and you think that the information is valuable, then request permission to restart. It is ultimately the decision of the participant to approve this.

Taking notes

The last piece of practical advice I provide is on note-taking. I advocate the practice of keeping notes during an encounter, as it can be extremely valuable in prompting further lines of enquiry as well as highlighting key points or ideas to explore further. Notes can be made at any time during an encounter, and you might want to establish a method for approaching this. Some might jot down information as participants speak, while others wait until a natural pause in the conversation to write things down. You might have specific approaches to taking notes, writing either in longhand sentences, short-form bullet points or using keywords, abbreviations or symbols. The method of recording notes can also differ. Some prefer handwriting notes into a book or pad, while others use Post-it Notes or cue cards. Technology, such as smartphones, tablets or laptops, can also be useful in recording notes; however, always make sure your participants know that you are using the technology for research purposes and not to check your social media!

There are, of course, common pitfalls with note-taking that are important to consider. It is perhaps common to assume that note-taking is the same as voice recording in that they both function as a way of recording information. In many ways, this is true, but the application of taking notes can, in some cases, be a source of anxiety for participants. From a participant's perspective, note-taking can sometimes come across as a secretive or exclusive practice, particularly if you are trying to be discrete. Suddenly writing something down while a participant is talking might be quite distracting if they are recounting an experience, and they could lose their train of thought or stop talking altogether. Being explicit with participants from the outset about if and how you will take notes and what types of things you will be noting down is good practice and helps ensure the participant is involved in all aspects of the encounter. The key message here is to take notes when necessary and to do this sparingly.

Building rapport

Rapport is necessary to varying degrees, but it can very often be difficult to get right every time. There is something extremely intimate about the interview process in terms of asking participants to reveal personal information about themselves to a stranger in a one-to-one situation. As an interviewer, it can be tempting to attempt to be objective and step back from the encounter, but as I discussed in Chapter 4, you, as an interviewer, are as much a part of the interview relationship as your participants. Building rapport between you and your participant can be valuable in encouraging fuller, more honest responses to questions. This might not be essential in every interview encounter, but as Valentine (2005) suggests, being reciprocal in sharing information about yourself with your participants humanises you as a researcher but also, in the context of research that you may be directly linked to (e.g., through your age, your employment status or skillset etc.), there is the potential for revealing yourself as someone who might share similar perceptions, experiences or values as your participants. Valentine (ibid.) does, of course, caution against revealing too much information. It would be unprofessional to spend time dwelling on your perceptions or relating your participants' responses to your own beliefs. Moreover, expressing your own views or opinions on the research topics may even lead the direction of the encounter, resulting in participants only telling you what they think you want to hear. Hence, transforming an interview into a full 'we' relationship can raise ambiguities as to whose experiences are being discussed and whose knowledge is produced from this (Seidman, 2006). It is important, therefore, to think carefully about how much of 'you' is relevant to the discussion being undertaken.

The art of listening

Listening might seem like an obvious point, but listening, not talking, is an interviewer's more valuable skill. Interviewers, by dint of their interest in the phenomenon under investigation, can often get very excited about the research being conducted. I remember transcribing some early interviews of my own and cringing at the instances where I had interrupted or spoken over a participant, particularly if it was during a critical point of their response. Indeed, Talmage (2012) suggests that interrupting participants is rarely ever helpful to the flow of conversation, so being quiet and staying focussed is crucial to effective interview practice. If you are using a voice recorder, it is tempting to view the technology as *doing* the listening for you during an interview. In some respects, this is correct, yet interviews do require considerable amounts of listening in order to ensure that you, as the researcher, understand the responses being given and whether these convey the requisite amount of information and detail needed for later analysis.

But listening is not just about following the content of the participants' responses. While what they say will ultimately be what makes it into the analysis, how participants respond, the types of words or phrases they use, and the inflection of their voices can reveal a lot to an interviewer about their motivations and intentions. As an interviewer, you are also going to be expected to concentrate on the direction of the interview. It is very easy to become absorbed by the responses and quickly lose the sequence of questions or the thread of a theme. Adopting a more active listening approach will ensure you are able to follow all registers in the interview. Active listening is often referred to as 'listening on purpose' and involves listening to your participant's speech while being attentive to non-verbal cues (both from your participant and your own) and responding empathetically (Weger et al., 2010).

Pauses and silences

Pauses or silences within an interview can feel extremely awkward when you are 'in the moment'. As an interviewer, you might view a pause as unresponsiveness or an unwillingness to respond to a question. In reality, pauses or silences are important opportunities for participants to think and reflect on the question being posed and formulate their response (see Box 6.2). Pausing is therefore as important a part of language as the words we convey. For example, the playwright Harold Pinter was famous for including significant pauses in his plays in order to replicate the natural silences of our everyday speech. What seems like a long gap in the interview setting is likely to constitute only a few seconds in the recording (Owens, 2006).

Box 6.2 Giving voice to silence in interview research

What is silence? Bengtsson and Fynbo (2018) posit that often, in research contexts, we consider silence to be negative, a form of disruption to the flow of an interview that affects the quality of interview responses and presents 'data gaps' in interview transcripts. Hence, in the context of a 'successful' interview, silences are usually considered something to be avoided. Yet, Bengtsson and Fynbo argue that silence is closely linked to power (see Chapter 4) and may be expressed in order to wield, challenge or resist power exchanges within interview encounters. This sounds rather problematic, yet the context, placement, duration and pattern of silences can provide important characteristics to an interview encounter that speech may sometimes miss.

Following Bengtsson and Fynbo (2018), we might consider silence in three intersecting ways:

Silence as protest – the act of disrupting the flow of an interview in such ways that expose unequal power relations (e.g., if you are perhaps talking too much or interrupting an interview participant to the point at which they stop talking).

Silence as rapport building – providing space for an interview participant to express their perceptions and experiences in their own way and own time (e.g., asking a question and then allowing the participant to answer it fully in their own time without interruption).

Silence as recognition of emotional empathy – pausing to allow the emphasis of an account or reaction to be noted (e.g., being attentive to the ways in which participants are responding to a question and not hurrying on to the next question).

So, rather than '*dealing* with silence', Bengtsson and Fynbo (2018: 33) encourage interviewers to consider utilising silence as a valuable and dynamic tool. They argue,

> This relationship contributes very diverse silences. Sometimes silence controls the interview situation; sometimes it turns the situation upside down. Sometimes silence played strategically right, and sometimes as mere intuition, produces rich data. [...] Silence constitutes possibilities for interview participants and interviewers to handle the complex power at play in qualitative interviewing either by maintaining or by losing control of the situation.

The role of non-verbal cues

So, it may already appear like you are going to have to concentrate on much more than just the words your participants convey in an interview, and there is yet still one more dimension that I want to raise. Non-verbal information – be this through body language, facial expression or registers like laughter, sighing or crying – is a highly important signifier of how a participant recounts the experiences you are asking about, as well as how they are responding to your questioning (Shuy, 2003; Fielding and Thomas, 2008). A participant might, for example, laugh when answering a question, and you might want to explore why they laughed. Was this a sarcastic response? Did they genuinely find something funny? Or are they laughing nervously frequently throughout the interview? It might be possible to ask for clarification during the interview, but this will need to be handled very carefully so as not to come across as rude.

Moreover, body language and facial expressions can provide very interesting ways of examining how a participant conveys information. There may be instances where the verbal and non-verbal do not match – i.e., the participant might be saying something positive, but their arms are tightly folded. This may be an expression of nervousness on the part of the interview participant, but it might also tell you something about the comfort of the participant in answering your question. You may need to consider whether what you are asking is potentially sensitive or if the participant thinks they are being led into responding in a certain way. As with verbal cues, it may not necessarily be possible, or appropriate, to challenge an interview participant about body language; however, if you think it might affect the quality of the data being produced during the rest of the interview, you could pause the interview and ask if they are comfortable before restarting.

Box 6.3   Common pitfalls in interview research

In their paper on learning how to interview in the social sciences, Roulston et al. (2003) presented analysis from discussions with novice interview researchers on their experiences of interviewing and the challenges they considered important in the interview research. Many of these have already been outlined in this chapter, and I draw them together here by way of listing some of the common pitfalls when interviewing and what strategies can be adopted to mitigate them.

Roulston et al.'s (2003) participants cited the following:

Dealing with unexpected behaviours from participants, including tardiness, background noise or distractions and incongruous behaviours – such as eating. Interviewers found that these actions often caught them off guard, leading to mistakes, fumbles and a general feeling of disorganisation.

Questioning their (interviewer's) ability to listen carefully. These included forgetting to inform participants of aspects of the interview (e.g., note-taking), (un)intentionally posing leading questions or making assumptions. Common among the participants was a need for them to acknowledge their interview approach (i.e., relaxed and conversational vs. formal and structured) from the outset so as to ensure that data being produced aligned with the research design.

Designing and delivering questions pertaining to the structure, form, flow and quality of interview questions. Going 'off topic' and wasting time were common pitfalls identified by the interviewers, and they reflect that getting to the point and sticking to it helps ameliorate any lack of focus.

Handling sensitive research topics. Both dealing with emotional responses to questions or potential hostilities if difficult or sensitive questions came out of the blue were common responses among the interviewers. In both instances, reflecting on and adapting interview practice is an important step in developing clearer strategies for interviewing, as well as being empathetic in order to build rapport.

Interference and interruptions

I begin this section by discussing the relative environmental challenges associated with conducting interviews, from the distractions of noise interferences through to the more elemental obstructions from the weather. Interviews are impossible to conduct in a vacuum in that you cannot remove all unwanted interruptions, but there are ways of identifying and managing potential obstructions so as to ensure conversations run smoothly and that you have good quality recordings to work with.

Unintended sound, what we might determine negatively as 'background noise', can be deemed disruptive in interview encounters, either interfering with the quality of the voice recording or distracting the flow of the interview itself. Hall et al. (2008) argue that researchers often go to great lengths to reduce the influence of noise on interview encounters. We might, for example, close windows, display 'do not disturb' signs on doors or turn off mobile phones or other electronic devices that might buzz, ping or chime at an inopportune moment. In these circumstances, we are, not entirely unreasonably, privileging an interview participant's voice in the research encounter, particularly as we want to make sure that interview participants are comfortable and able to concentrate on the task at hand during the interview and that we also ensure our interview participants' words are accurate and legible in later transcripts. Preparation is crucial here in avoiding unwanted noise interruptions. If you are using a particular location for all your interviews (e.g., a private office or room in a workplace, school or university, a room booked in a library, or a convenient public space like a coffee shop or café), then take time to scope out the environment. Observe whether there are any points of the day where the location might be very busy or might be susceptible to noise. Take note of any occurrences that might disturb the interview, such as scheduled fire alarms, that might interfere with the interview. You clearly cannot mitigate against all noise, but taking steps to identify probable interferences may help you minimise their impact.

Data quality is of primary concern with all interviews, and as I advised earlier in this chapter, it is important that you check your equipment and the location you are interviewing in prior to each encounter. If you are able to control any background interferences, then do so, but some interruptions might not be so easy to mitigate, such as fire alarms, chatter in coffee shops or roadworks. In these instances, being aware of the potential for interruption and having a protocol for handling this is important. This might include letting your participants know what to do if something unexpected were to happen (i.e., that the interview might need to be paused or even relocated) should help keep everything flowing as much as possible.

Alongside environmental and ambient interruptions, other people not connected to the research can also be a hindrance to the success of an interview. This might be as simple as a noisy group of people sitting adjacent to you (in which case, you might easily consider moving the interview elsewhere) to more extreme cases, such as unwanted interactions with strangers who involve themselves in your encounter. Like environmental noise, there are steps you can take to mitigate against interruptions from people, such as interviewing in private spaces or informing others around you that an interview is taking place. This is, of course, not easily achievable in public spaces, and you might need to think carefully about the probability of an interruption taking place and what impact this might have on the success and quality of the interview. MacDonald and Greggans (2008) talk of the complexities of interviewing families with very young children and the perceived 'chaos' that can ensue when interviews are interrupted and paused in order to handle an ever-changing environment. Moreover, if you are, perhaps, interviewing a participant in their workplace, you might run the risk of interruptions from colleagues. In any of these scenarios, the most effective approach is to pause the interview (both the conversation *and* the voice recorder), wait for the interaction to finish, and then restart again with permission from your participant. Crucially, remain professional throughout and avoid confronting the interrupter. If the interruption persists, then stop the interview and find a new location or propose rescheduling it if the interview is disrupted.

Safety

Alongside interruptions, safety is an important consideration in interview practice. I explained the rationale behind identifying and mitigating against risk in Chapter 5 and it is important to think carefully about the relative safety implications associated with interview encounters while they are being conducted. These are not exhaustive, and there are some excellent 'safety in the field' guides for interview encounters that you may find useful.[2] Wherever you conduct your interviews, always ensure you remain fully aware of your surroundings. This includes getting to and from an interview, so employ basic safety when crossing roads, using public transport or driving a private vehicle. Try not to have valuable items, such as smartphones or tablets, on show while interviewing. If they are necessary for recording the conversation – put them away while they are not in use. If you or your participant feel unsafe at any point during an interview, particularly if it is being held in a public place, then please terminate the interview and leave immediately. Moreover, while interviewing ostensibly invites participants to disclose information about their lives, it is vitally important to avoid documenting anything deemed illegal or dangerous. From your perspective as an interviewer, this extends to being mindful of permissions to interview in certain locations or to record or document sensitive information.

What to do if things go wrong

This chapter is, perhaps, beginning to sound rather gloomy. It is, of course, not my intention to scare you, but anticipating some of the problems that can occur during interview encounters should make your interview practice enjoyable, relaxing and productive. In this section, I cover two of the more pragmatic problems that can arise during interviews – technical issues and dealing with distress, hostility or emergencies.

Technical issues

A crucial step in successful interview practice is to check and test your equipment. If you are using technology for the first time, such as a brand-new voice recorder or a recently downloaded voicenote app on your phone, make sure you understand how to operate it and have access to any instruction manuals/troubleshooting guides during the interview itself. Always check that equipment is fully charged and that you have spare batteries for anything battery-operated. Moreover, check how and where your device stores audio files. If it holds them on the device, then make sure you have enough storage capacity for the interview you are conducting. This might involve uploading existing files from your device onto a computer to free up room or adding external storage, like memory cards, if your device supports these. Get used to uploading audio files to a secure storage space as soon as you have completed an interview. Remember that you only really have one shot at getting an interview with a participant, and it may be difficult, or even impossible, to locate or retrieve lost or overwritten files. Finally, to avoid poor-quality interviews, test the audio quality before each interview to ensure both you and your participant are clearly audible. This will allow you to make adjustments to the location, your positioning and the placement of the voice recorder ahead of the interview commencing.

Distress, hostility and emergencies

Beyond technical issues, there can be circumstances in which unanticipated problems arise between you and the interview participant, such as participants experiencing distress or becoming hostile, or an emergency occurring – be it medical or security – that you need to deal with 'in the moment'. Please note that significant issues are rare but remain possible, so it is important to be aware of the potential for something unexpected to occur and to have a strategy and contingencies in place ahead of the interviews taking place.

I have outlined approaches to mitigating hostility and distress elsewhere in this book in terms of designing interview themes and questions and the use of piloting to test the robustness and sensitivity of questions (Chapter 3). There may, however, be instances where sensitivities are not known until questions are presented to a participant, and it may well be that a particular question, however sensitively it may be designed, could elicit a traumatic, upsetting or angry response from a participant. One of the best strategies to minimise the risk of harm is to provide interview participants with a list of the themes you intend to ask or to highlight any potentially sensitive questions prior to the interview. This is a good opportunity for participants to either prepare to answer something sensitive (be it recalling a distressing past experience or providing an opinion on a sensitive topic) or inform you of any questions they would not be happy responding to. This should not be treated as a given, though, and in the heat of the moment, a participant might feel unable or unwilling to respond. Your participants, of course, have the right not to answer a question, so it is good practice to build warnings into the interview when sensitive topics are to be raised.

If issues do arise and a participant becomes distressed or hostile, the interview must be either paused or terminated (including turning off the voice recorder) and the interview participant given the opportunity to temporarily step out of the interview or leave altogether. If the participant has requested a moment to compose themselves, then allow them space and time to return, and ask if there are any changes in the interview approach they might like you to take (e.g., the pace of questioning, the removal of certain themes or topics or to reconfigure the set-up of the interview space). If the interview participant is unable or unwilling to continue with the interview, then you must respect their right to withdraw, including not using any recorded information as data in your research. Always have information at hand about counselling and support services in the area local to your participants and pass this on to the interview participant at the earliest convenience.

SUMMARY POINTS

- Consider the basics of interview practice, such as careful recruitment and scheduling, getting to grips with recording and taking notes.
- Remember that interview participants are human beings, so developing rapport, listening, avoiding interruption and watching out for non-verbal cues will help the conversation flow smoothly and comfortably.
- Plan for any interferences – environmental or human – and consider mitigation strategies.
- Think carefully about safety, technical issues and how to manage distress, hostility or emergencies.

WHAT TO DO NEXT?

- Crucially, have fun doing your interviews!
- Follow the aforementioned advice on mitigation strategies. If in doubt, pilot your interviews again to make sure you are comfortable and prepared.

Suggested further reading

Bengtsson, T. T., and Fynbo, L. (2018). Analysing the significance of silence in qualitative interviewing: Questioning and shifting power relations. *Qualitative Research*, *18*(1), 19–35.
    Putting silence as the central focus of this article, these authors provide an excellent account of how and why researchers must be aware of their own, and their participants', practice during interview encounters.

Brinkmann, S., and Kvale, S. (2018). *Doing interviews*. London: SAGE.
    Chapter 5 of Brinkmann and Kvale's book provides some excellent advice on putting interviews into practice.

MacDonald, K., and Greggans, A. (2008). Dealing with chaos and complexity: The reality of interviewing children and families in their own homes. *Journal of clinical nursing*, *17*(23), 3123–3130.

This paper explains the complexities of handling interruptions and provides advice on when interruptions are okay and when they are not.

Talmage, J. (2012). Listening to, and for, the research interview. In Gubrium, J. F., Holstein, J. A., Marvasti, A. B., and McKinney, K. D. (Eds), *The SAGE handbook of interview research: the complexity of the craft* (2nd ed, pp. 295–304). London: SAGE.

In this chapter, Talmage provides some supportive commentary on developing meaningful listening strategies.

NOTES

1   It is also important not presume that all participants will consent to being recorded, and there may be instances where an interviewer may have to rely on taking detailed notes.

2   See the University of Oxford's (2023) "Staying Safe When You Are Interviewing Guide": chrome-extension://efaidnbmnnnibpcajpcglclefi ndmkaj/https://socsci.web.ox.ac.uk/files/safeinterviewingremindersjan 19pdf

REFERENCES

Bengtsson, T. T., and Fynbo, L. (2018). Analysing the significance of silence in qualitative interviewing: Questioning and shifting power relations. *Qualitative Research*, *18*(1), 19–35.

Brinkmann, S., and Kvale, S. (2018). *Doing interviews*. London: SAGE.

Fielding, N., and Thomas, H. (2008). Qualitative interviewing. In: Gilbert N (Ed.) *Researching social life* (3rd ed) (pp. 245–265). London: SAGE.

Hall, T., Lashua, B., and Coffey, A. (2008). Sound and the everyday in qualitative research. *Qualitative Inquiry*, *14*(6), 1019–1040.

Hayes, T. A. (2021). The practicalities of researching creatively. In, von Benzon, N., Holton, M., Wilkinson, C., and Wilkinson, S. (Eds). *Creative methods for human geographers*, (pp. 61–72) London: SAGE.

MacDonald, K., and Greggans, A. (2008). Dealing with chaos and complexity: The reality of interviewing children and families in their own homes. *Journal of Clinical Nursing*, *17*(23), 3123–3130.

Owens, E. (2006). Conversational space and participant shame in interviewing. *Qualitative Inquiry*, *12*(6), 1160–1179.

Roulston, K., DeMarrais, K., and Lewis, J. B. (2003). Learning to interview in the social sciences. *Qualitative inquiry, 9*(4), 643–668.

Seidman, I. (2006). *Interviewing as qualitative research: A guide for researchers in education and the social sciences* (3rd ed). New York, Teachers College Press.

Shuy, R. W. (2003). In-person versus telephone interviewing. In: Holstein, J. A. and Gubrium, J. F. (Eds). *Inside interviewing: New lenses, new concerns* (pp. 175–193). Thousand Oaks: SAGE.

Talmage, J. (2012). Listening to, and for, the research interview. In Gubrium, J. F., Holstein, J. A., Marvasti, A. B., and McKinney, K. D. (Eds), *The SAGE handbook of interview research: The complexity of the craft* (2nd ed), (p.p. 295–304. London: SAGE.

Valentine, G. (2005). Tell me about… Using interviews as a research methodology. In, Flowerdew, R. and Martin, D. (Eds) *Methods in human geography: A guide for students doing a research project*. Harlow: Pearson Education Ltd.

Weger Jr, H., Castle, G. R., and Emmett, M. C. (2010) Active listening in peer interviews: The influence of message paraphrasing on perceptions of listening skill, *The International Journal of Listening, 24*(1), 34–49.

7

USING NON-VERBAL MATERIALS IN INTERVIEW PRACTICE

Chapter objectives

This chapter will examine the ways in which interview practice can be enhanced using non-verbal materials, such as photographs, objects, sound and video. By reading this chapter, you should

- understand when it is appropriate to include non-verbal materials in interview practice,
- recognise the difference between 'found' materials and those produced by participants,
- consider strategies for including 'talking while doing' activities in interview encounters.

INTRODUCTION

You may recall that in the opening sentence of Chapter 1, I outlined interviews as a 'conversation with a purpose', and in many ways, this entire book is about how this conversation is managed. Yet, conversations might not just be about talking, and beyond 'the voice', I want to examine how other non-verbal mediums, such as visual materials. objects, sound and video, can be useful in eliciting particular forms of knowledge from interviews. In most conventional interview approaches, the questions you ask your participants will produce excellent responses; however, there may be some circumstances where participants might need to be prompted to respond to a specific piece of information or provide evidence of their engagement with an activity or process in preparation for the interview. Non-verbal materials

DOI: 10.4324/9781003292784-9

can be a very intriguing way of encouraging deeper, more meaningful discussions with participants in ways that potentially transport them to different spaces and times. Throughout this chapter, I explore the ways that images (be they static or video-recorded), objects and sounds can be used to help augment an interview encounter. These might be user-generated content (i.e., produced by the interview participant), found by either the researcher or interview participant or materials produced or gathered during an interview by participants.

In order to understand their value in interview research, we must first consider what counts as non-verbal materials. Visual images, for example, can come in different forms, including static materials such as maps, graphs, charts and photographs or moving images like video or film). Objects can be almost anything, from works of art right down to promotional flyers for an interview participant's favourite music artist. It is often the latter objects – ephemera, memorabilia and tactile 'stuff' that we engage with in our everyday lives – that can be most memorable, desirable or illuminating during interviews. This might include anything from postcards, board games or badges (Merriman, 2005) to historic household items and inventories (Evans, 2008). Non-verbal materials need not be static, and sound and video can be equally as productive in supporting interview practice. Your participants might watch excerpts from a film or even produce their own films that are then discussed in an interview. Moreover, you might accompany them on a 'sound walk' or ask them to listen to a piece of music. So, while this chapter steps slightly away from 'the basics', I will demonstrate ways in which non-verbal materials can add value to interview practice, using a variety of creative and innovative approaches that you can incorporate into your own research (see Rose, 2016; von Benzon et al., 2021).

In outlining the basics of non-verbal interview materials, first, I discuss the ways in which found and participant-produced images can be built into interview practice. Next, I examine the role of objects and 'talking while doing' in supporting interview practice for sensitive topics. Finally, I switch the focus to interviewing about, and with, sound and video.

USING AND PRODUCING IMAGES

I start here with visual materials – mediums that have become extremely popular among social researchers when it comes to finding new and creative ways of interviewing participant groups (Glegg, 2019; Woodward, 2016; Henwood et al., 2018). While there are many ways that this can be achieved, interviewing with images can be divided into two distinct categories. First, images might be existing or 'found'. These are materials that have been produced by others external to the project and for a different purpose and then collected by the researcher or participants for the study. Examples of found images might be advertising campaigns, flyers, movie posters or album covers. Second are images that are produced by the researcher or participant for the purpose of the research project. These are what can be referred to as 'intentionally produced' materials as they fit directly within the context of the project. In the following sections, I will explain how to use found and participant-produced images, and at the end of these sections, I will provide some general guidance for incorporating any images into interview practice.

'FOUND' IMAGES

Interviewing with found images has increased in popularity in recent years, partly because they act as prompts for conversation (excellent if you are concerned your interviews might dry up), as well as being useful for corroborating points raised by participants (Felstead et al., 2004). Building found images into interview practice commonly involves the researcher gathering together a range of images that speak to the research themes under investigation and then asking participants to provide commentary, opinions and reflections on them during an interview. Found images might also involve participants being asked to compile a body of their own materials that relate to their interpretations of the phenomenon you are investigating. You might provide a set of instructions on the types of images you would like participants to collect or direct them to a particular website, gallery or publication to select their own images from. The images collected will then inform themes to be discussed during an interview that relate to where the images came from, why they have been selected and what they tell us about the issue (Rose, 2016).

There are obvious benefits to incorporating found images into interview design. These types of materials are usually easily collected and can be widely available and accessible to researchers and participants in terms of quantity and scope, meaning a 'representative' sample of images per participant or project is achievable. Found images can also allow for deeper interrogation of a phenomenon as they encourage the interviewer and interview participant to critically examine them together – for example, you might focus in on particular parts of an image during the interview or ask your interview participant for an explanation of what the image means to them. Moreover, there is significant value in questioning your participants for commentary and evaluation on the production of images (e.g., who produced it and what for) and how images are consumed (e.g., who the expected audience is).

That said, incorporating found images comes with challenges, and there are things to consider before adopting them into interview practice. You will need to find a way of demonstrating that each image source is credible and that the images themselves are accredited to whoever produced them (e.g., checking whether you have permission to share found images or replicate images into your write-up). Moreover, it is not enough just to include visual materials for the sake of it. As with all methods, make sure you justify the inclusion of the method and analysis technique and document how and why they have been used.

Box 7.1   Activity – Using found images in an interview

You are studying the topic "Imagining and responding to climate change" and want to interview people about their perceptions and experiences of the phenomena. You intend to incorporate images of climate change events from around the world into your interviews to find out how your participants might feel about and relate to climate change impacts.

Consider the following questions:

- Where can these images be found (e.g., online, in magazines or newspapers, in advertising campaigns)?
- What format might they take (e.g., photographs, leaflets, adverts, promotional materials)?
- How would you incorporate them into the interview (e.g., provided in advance, presented during an interview, grouped together, shown one by one)?

- How would you record your participants' engagement with these images (e.g., verbally using the voice recorder, in a separate note-book, by annotating the images)?

Once you have worked through these questions, consider what benefits might come from adopting your method, what issues might need to be considered and what strategies you could take to mitigate these issues.

PARTICIPANT-PRODUCED IMAGES

The alternative to found images are those produced by the researcher or participant. Researchers might seek to document a certain phenomenon and then use the images that have been produced by themselves as the basis of a set of interviews (see Clark-Ibáñez's (2004) research with young children in Los Angeles, United States, and Folkestad's (2000) work on Norwegian adults with 'intellectual disabilities' preparing to re-integrate into local community housing). More common, though, are participant-produced images, and these might take the form of photographs, sketches or maps. One such method is called 'photovoice' (see Wang, 1999; Strack et al., 2004), whereby participants take photographs relating to a place, issue or experience and then work with the researcher to examine and explain the images produced in follow-up interviews. Participant-produced images therefore involve the participant being given some form of brief or set of instructions pertaining to how the images need to be taken, what could be included and any guidance on things like framing or what should not be included in the images. Crucially, as they are produced by the participant, images might be created in advance of an interview, so the researcher can run some very broad thematic analysis that helps guide the interview, or they could be produced during the interview encounter itself, with the interviewer asking questions about what is being produced and why this matters to the participant.

This type of approach is very popular in qualitative research as participant-produced images have less interference from the researcher, making the later interview more about 'their' experiences of producing the images rather than interpreting the potential intentions of existing materials. Hence, participant-produced images have significant benefits to interview research design as they align closely with the interviewing ethos of researching 'with' participants rather than 'about' them. This can also be a popular method for conducting research with vulnerable groups, such as children, who might feel intimidated by a classic face-to-face interview with an adult.

As with found images, there are challenges involved in incorporating participant-produced images into interviews. Images are, of course, not inert and have significant values attached to them (Rose, 2004, 2016). The identities of the image maker are important to consider, particularly in terms of how the image is framed (i.e., what has been included or left out from the boundaries of the images) and the relative importance of the image's subject matter to the creator. Moreover, there can be ethical considerations around anonymity and confidentiality. Some researchers request that participants avoid including people's faces in images, while others edit the images to remove or blur out identifying information. There can also be a significant pitfall in not really analysing the images alongside the interviews themselves, with some researchers instead using images to simply illustrate points in the writing. Asking participants to produce images should be a meaningful exercise that involves drawing upon them in the follow-up interview to help interrogate the phenomenon they have captured (see Box 7.2).

Box 7.2   Activity – Building participant-produced images into an interview

You are conducting a research project on young people's engagement with public space. You want your participants to take photographs of the everyday spaces that matter to them and reflect on their routines and behaviours within them. You will use these images during the interviews to help lead the conversation.

Consider the following questions:

- What would you ask your participants to produce images of and why?
- What guidance should you give your participants for taking their own photographs?
- Are there things that your participants should avoid including in their images?

> • How will you use the images as part of the interview encounter to help understand their experiences and perceptions?
>
> Once you have worked through these questions, consider what benefits might come from adopting your method, what issues might need to be considered and what strategies you could take to mitigate these issues.

How to 'do' interviews with images

In this section, I list some of the common things to consider when using images in your own research practice.

1. Provide clear briefing instructions. As with any interview, clear instructions are important; however, asking participants to include or work with images will require you to make it absolutely clear what is expected of them. Consider outlining the following:

   a. What the images are for
   b. What you are asking participants to do in preparation for, and during, the interview
   c. Any specific advice on what is to be produced
   d. How the images will be used during the interview
   e. What happens to the images after the interview

2. Think about how you will use the images during the interview. Will they be displayed on a screen (e.g., a laptop, monitor, television, tablet, phone) or in physical form (e.g., printed photographs, pages cut from a magazine, art hung on a wall)?

3. Think also about how they might be interacted with (e.g., viewed from a distance, handled by the participant, annotated by the interviewer/interview participant).

4. Be realistic about the volume of images you can handle during an interview. Discussing seven to ten images during an interview will usually take up to one hour, so think carefully about the value of adding more to the encounter.

5. As above, consider the timing and pacing of the images. Are you grouping them together into themes, or do you want to discuss each image one by one? Will you need to allocate a rough time limit for each image to be discussed? Are you choosing the

order to work through the images, or will you let your participants decide on this?

6. Finally, try to make sure that your interview participants have the ability to add their own thoughts and opinions on the images selected during the interview. Provide opportunities to ask if there is anything they want to add or allow time to reflect on the images.

What to talk about

Moving beyond the images themselves, it is important to consider carefully how you will use them during an interview encounter. This sounds somewhat obvious, but it is possible to get caught up in developing the method and neglect the interview approach itself. Think carefully about what you want to get out of using images in your interviews. If you are asking participants to recall information contained visually in an image, then having that image to hand will make it easier to manage the process. You might start by asking your participants to talk about why they chose the image, or if it is one you selected, what they feel about it. Starting broad and general can help a participant relax into the process. Once you have both shared some general thoughts and ideas, move on to specific elements of the image. There may be details of a photograph that you want to know more about, so focus attention on those. Likewise, to avoid any visual bias from you as the researcher, you might ask participants to pick out any details that matter to them. If your participant has taken the photograph, consider the framing of the image to explore both what is included in the photograph and what is left out and whether the image was framed consciously or sub-consciously. Finally, the elements of the images you discuss in the interview are likely to trigger a broader conversation about things perhaps indirectly related to the photographs. For example, if you have asked your participants to provide photographs of their favourite holiday destination, you might inevitably talk about how this relates to other destinations they have visited and whether they chose the trip because of the activities depicted in their photographs.

OBJECT-ORIENTED INTERVIEWING

In this section, I extend the aforementioned examples of using images in interview practice to discuss the role of material objects in helping produce knowledge in interview encounters. Objects work in the same way as photographs in that they can prompt responses and encourage deeper engagement with a phenomenon; however, they have added benefits in that they can be tactile and in the context of interviews that require objects to be produced (e.g., crafted from clay or LEGO, or written down in poetry or music), active and participatory, in giving both participants and interviewers something else to concentrate upon during the interview (Dunne and Pimlott-Wilson, 2021). In this section, I outline two practices: 'interviewing with objects' and 'talking while doing'. Unlike working with images, these activities are very different from one another and will require different approaches to managing the interview process. I will include specific 'how to' advice within each section.

Interviewing with objects

In this section, I consider the value of using objects in interviews both in terms of triggering memory and providing props for further discussion. Peters (2011) explains that everyday items, such as tourist souvenirs, can be very useful in encouraging interview participants to reflect on past practices and consumption habits. During interviews conducted in participants' homes that examined the ways that tourist souvenirs might help us understand people's experiences of different places, Peters found that her participants often picked up and examined their souvenirs while they talked about them, using them as tools through which to transport back to the places in which they were obtained and provide deeper context of their meaning to their current everyday lives. Notwithstanding this, Peters found that, when asked if objects could be photographed, her participants often moved these objects away from the spaces they were displayed in their homes, somewhat decontextualising them from their meaning. We might, therefore, need to be a little critical of what objects signify in research contexts and how our participants interact with them during conversations.

An alternative way of incorporating objects into interviews is to consider how objects are engaged within 'normal' settings. In order to understand how expatriate British women living in Dubai, United Arab Emirates, developed a sense of belonging in new, temporary environments, Walsh (2006) spent time with her participants, observing their daily activities, such as domestic chores and leisure activities, and interviewed them about how they engaged with the objects present within their homes. Where Peters' (2011) participants decontextualised their objects when asked to discuss them, Walsh's provided deeper context through everyday practice. Hence, Walsh found this a productive way of understanding the intricacies of everyday life and what objects might mean to people. You might, therefore, find benefit from introducing objects into your interview research design, particularly if your research is about how people might engage with everyday items, like postcards, sports equipment, toys and games or technology.

Box 7.3   How to interview with objects

In this box, I provide some helpful advice on what to consider when using objects in interview practice.

1. As with images, always provide clear instructions as to how and why the objects will be incorporated into the interview preparation and practice.

2. Think practically in terms of risk. Are you expecting a participant to bring something valuable to an interview that might get lost or damaged? Might it be impractical to transport heavy, bulky or fragile items to interviews? Are your participants happy for you to pick up and handle their objects?

3. Use questions that will encourage interview participants to talk about the object (e.g., if you have asked an interview participant to bring along a cherished artefact from their childhood, build the questions around why they chose that item and the meaning behind it).

4. Avoid making presumptions about items chosen by participants. An object will mean something to them, so use your questioning to encourage their meanings and interpretations to come through.

5. Think about where the object, or objects, will be placed during the interview. Does an object need to be situated close enough to be picked up and interacted with, or is it better placed some distance away so as to be talked about? Also, consider whether the object will be required for the entire interview. If it is not, then perhaps put it away so as not to distract from the conversation.

6. Finally, think about how you will record the interactions with the object. If it is not yours, then you might need to photograph it. You might also need to take separate notes or even video the interview if you want to record the ways in which your participant handles the object.

'Talking while doing'

In contrast to using existing objects to support interviews, other approaches place value upon conducting activities, such as making things, during an interview encounter that can help produce certain types of, often hard to understand, knowledge. Research that involves participants talking while forming plasticine models, assembling LEGO, drawing pictures or building collages can help focus attention on the production of specific tasks or talk about difficult or sensitive subjects.

Eldén (2013), for example, has examined the ways in which children's voices might be articulated differently through the use of creative activities conducted while interviewing them. She conducted a 'draw-your-day' exercise whereby her young participants were asked to draw four activities that happened at predetermined times on a given day. The children were asked questions during the process about what they were drawing, who they were depicting and why they had chosen certain activities to draw. In this sense, the knowledge produced from this type of activity-based approach is co-produced between the interviewer and interview participant, meaning the interviewer has the opportunity to question what the interview participant thinks about a certain phenomenon – in Eldén's case, relationships of caregiving – visually as well as verbally. While Eldén, of course, cautions against considering this knowledge to be any more authentic than conventional interview approaches, she argues that the layers of information produced through 'talking while doing' capture very realistically the complexities and messiness of everyday life in ways that simply talking about it might obscure.

Box 7.4   Doing as making

In examining the ways in which we might incorporate 'doing' into interview practice, I draw on two contrasting examples based on the sharing of specific tasks and the articulation of sensitive subjects.

I start with Harrison and Ogden's (2019) research on crafting practices. They set up what they called 'knit "n" natter' sessions that involved bringing together groups of participants to perform knitting practice together while encouraging them to discuss their experiences and practices associated with knitting. In practical terms, this process involves the sharing of practice, in this case expert or specialist knowledge, among a set of like-minded individuals to help explore how participants might experience the practice in their everyday lives. The benefit of this approach is that the familiarity of the activity effectively "simulate[d] the openness and relaxed conversational flow of a social scenario" (ibid.: 458) and, in Harrison and Ogden's research, provided opportunities for discussion to emerge more fluidly than if, perhaps, the participants were interviewed 'about' their knitting practices.

In terms of sensitive subjects, Dean (2015) developed a critical drawing exercise whereby university students were asked to draw 'what homelessness looks like' and then take part in semi-structured interviews to discuss their drawings. Like Harrison and Ogden's (2019) account earlier, this example of doing while making allowed participants to relax into the exercise while at the same time connecting the research process to a meaningful activity and the knowledge that is produced through this type of engagement process. What makes this a valuable tool to explore sensitive subjects, though, is the ways in which drawings can reveal participants' internal representations of a phenomenon – in this case of homelessness – and provide a visual depiction of the phenomenon through which interviewers can prompt, question or clarify how and why the participants' have approached the task in the ways in which they have.

Common across both approaches, though, is a caution against presuming practices of talking while doing as producing superior knowledge. As Dean (2015: n.p., emphasis in original) argues, these methods "produce different knowledge differently", meaning they should not replace the conventional interview but, within the right context, be used to help support the acquisition of more robust knowledge.

In the remainder of this section, I will take you through a worked example of how a practical activity – in this case drawing a mind map – can be incorporated into interviewing, with some detailed steps on how this works in practice. Boden et al. (2019) developed a technique called 'relational mapping interviews' that incorporates user-generated content into interview practice. This approach involves participants being encouraged to individually produce visual 'mind maps' which document their experiences, perceptions and feelings of a phenomenon. Boden et al. (ibid.) suggest that this technique breaks the interview into discrete parts (or 'touch points'), which have specific tasks attached to them. These 'touch points' help guide the participant through the process and provide structure to the interview. During each task, participants are encouraged to effectively 'talk while they draw', with interviewers responding with follow-up questions as the images emerge. In Table 7.1 I have listed the four 'touch points' along with some explanation of how they can be used in practice and included suggested follow-up interview questions. The connections between doing and talking make

Table 7.1 Examining the four' touch points' associated with 'relational mapping interviews'

Touch point	Description of process	Follow-up questions
Mapping the self	This first step involves reflecting on who the participant is and their position in the world. Give participants the opportunity to do this how they wish, but provide suggestions if they struggle (e.g., they could draw a stick person, use words to annotate or symbols to represent things).	Ask questions that pick up on the details of the image. "Why did you decide to place the image in the centre of the paper?" "What motivated you to use that particular colour here?" "This is an interesting symbol. What does it mean to you?"
Mapping important others	Step two involves describing the things, people and places that matter and the relationships and networks between them. This might involve staggering the draw-talk-draw-talk process, perhaps if participants need reassurance that what they are producing aligns with the brief.	As above, choose questions that help the interviewee unpack their drawing. Also, consider looking for relationships and contrasts in the elements that are drawn: "This XXXX is interesting; tell me how it relates to YYYY?" Moreover, you might want to ask participants about things they have chosen not to depict. Always enquire if it is okay to talk about omissions, though, and be respectful if they do not wish to discuss them.

(Continued)

Table 7.1 (Continued)

Touch point	Description of process	Follow-up questions
Standing back	In step three, you can encourage the participant to appraise the entire map and provide commentary on the 'bigger picture'. This might involve physically 'stepping back' from the image to examine it in full, perhaps by sticking it to a wall.	This involves asking questions that encourage participants to reflect on what they have produced and to come up with their own meaning for it. "Now you have finished the picture, what do you think it looks like as a whole image?" "Is there anything about the picture that surprises you?"
Considering change	The final step involves making linkages between the past, present and future. This will likely mean reflecting back on the picture produced and considering what an ideal future, or outcome, could look like.	Ask questions that explore potential outcomes relating to the past or present issues, problems or occurrences depicted in the image. "In an ideal world, is there anything you would change about this map to make things better?" "You drew XXXX in your picture. Is there anything you could suggest to change this outcome?"

Source: Adapted from Boden et al., 2019

methods like relational mapping interviews effective approaches to 'get at' contentious and hard to vocalise feelings or perceptions in less-invasive ways than conventional interviewing.

INTERVIEWING WITH VIDEO AND SOUND

In this final section, I explore the potential benefits of including sound and video into interview practice. In some respects, these types of non-verbal materials operate in the same way as both still images and objects, and much of the advice I provide elsewhere in this chapter can also be related to interviewing with sound and video. There are some subtle differences though, as well as some ways of using these mediums in different ways to images and objects that I will outline in the following sections.

Video

Starting with video, there may be instances where you feel it necessary for your participants to engage with media derived from popular culture, such as cinematic films, television programmes, music videos, documentaries or advertising. This might involve you asking your participants to watch a piece of media ahead of an interview and then interviewing them about their experiences and thoughts in a follow-up interview. Banaji (2010) took this approach, conducting interviews with a variety of participants in Bombay, India, and the United States after screenings of the feature film *Slumdog Millionaire* to find out how participants considered Indian culture to be produced and represented through popular culture. This can be a valuable way of exploring opinions, or expressions of emotion, relating to popular culture and can help develop an understanding of how different demographic groups might relate to certain types of media. In a practical sense, this usually involves the participants watching the media ahead of the interview in their own time, so it is important to provide enough time for them to prepare for this. You will also need to ensure they are able to view the media in an equitable format. It would be unreasonable for your participants to have to pay to watch the media for your project or to have to download software they otherwise would not want to engage with. Moreover, if you are sharing the media, make sure you have the correct permissions to do so legally.

Another option might be to use video during the interview itself (Smith and Dunkley, 2021). You might have a set of clips from a film or television programme or a collection of adverts or music videos that you could show at intervals during the encounter and then explore opinions and feelings about these clips using follow-up questions. This can be useful in capturing reactions 'in the moment' and may avoid instances where participants might have forgotten what the video was about, particularly if they were not fully focussed on it at the time. Showing video during an interview has, of course, some drawbacks, and you will want to brief your participants very clearly ahead of the encounter. Avoid showing videos that might cause distress to your participants. If there is potential for the video to cause harm, then explain the content ahead of it being shown, and even consider providing it before the interview takes place. Always respect the wishes of your participants if they choose not to view anything. Using video content during an interview will require you to choose a suitable location for it to be viewed. You might require a television screen, for example, or use a laptop or tablet. You will also need privacy to ensure that participants can view and hear the video content while not feeling self-conscious about watching it. Avoid public spaces where you might be overlooked, but if this is not possible, then consider using headphones to control the volume of the video.

Sound

I deliberately separate sound from video here, as sound can be quite abstract in the medium through which it is produced and the ways in which sound is interpreted. There are differences, for example, in asking participants to relate to the lyrics and tune of a song and encouraging reflection on the way the sounds of a local woodland make them feel about the environment (Whittaker and Peters, 2021a, 2021b). They are likely to require different approaches to performing and recording the sounds (e.g., pre-recorded vs. live sounds), the types of questions asked and the structure of the interview itself (e.g., questions asked after the sounds have been performed or posed during the live event). Moreover, you could even consider working with your participants to produce sound, such as music or poetry, during the interview itself and discuss the processes involved in creating it. Kelly (2015), for example, developed an innovative 'audio documentary' methodology that involved young people experiencing homelessness meeting in a music studio over the course of 12 sessions and engaging with the technology to produce an audio documentary that exemplified their experiences of homelessness and of being in the music studio. Throughout the process, the participants were interviewed about their experiences and the sounds that were produced through the audio documentary provided opportunities for participants to express their feelings and emotions in non-verbal ways, with opportunities to elaborate on these in the discussion.

Hall et al. (2008: 1030) argue that working *with* sound can be profitable for understanding how knowledge is produced in an interview. They use the term 'soundwalks' to imply the ways in which interviewers might infer "meaning, feelings and associations triggered by [environmental sounds]" and how these can link to, or contradict, the discussion elicited by the interview participant.

Moreover, Gallagher (2015) sought to actively draw everyday sounds into his interview practice, using a combination of field audio and interview recordings to help provide a richer context for the research he carried out. Adopting an 'audio walk' approach, Gallagher (2015) attempted to layer the voices of participants under and over the soundtrack of the spaces in which the encounters took place, such as the sound of rain over the top of someone's speech. We might, therefore, infer from this then that while an interview is effectively a story that is narrated to us by our interview participants, sounds can provide interviewers a supplementary way to understand more about how these stories are produced and why they matter.

In a practical sense, like employing video, using audio will require some additional technology to either play or record the sound involved in the interview encounter. This might mean that you need to consider whether the audio you are including in the interview is to be recorded alongside, or separately from, the interview conversation. Think about whether it will be possible to analyse both 'layers' simultaneously. There might be certain instances in which the audio (for instance, music or atmospheric sounds) drowns out or interferes with the speech on the recording. Conversely, if you separate out the audio from the interview narrative (for example, pausing the recording to play a song), might it be difficult to match these together later when analysing? In these circumstances, it is always important to pilot the interview to ensure that the integrity of the data is not lost (see Chapter 3).

SUMMARY POINTS

- Interview practice can be enhanced using non-verbal materials, such as photographs, objects, sound and video.
- Photographs can provide context for an interview encounter and act as a powerful memory prop for recounting information.
- Objects are tactile and can be useful if interview participants are nervous or uncomfortable with being interviewed. This extends to 'talking while doing', which can encourage interview participants to focus on specific tasks.
- Sound and video can be incorporated within interview encounters to find out about engagements with anything from popular culture to local environments, as well as the everyday lives our participants inhabit.

WHAT TO DO NEXT

• Might there be situations whereby your interview could be enhanced by using non-verbal materials?
• If incorporating material 'things' into research (e.g., photographs, video, objects), think carefully about how you will incorporate them into an encounter, whether they will be introduced by you or the participants, and how you intend to record their presence in the interview.

Suggested further reading

Felstead, A., Jewson, N., and Walters, S. (2004). Images, interviews and interpretations: Making connections in visual research. In Pole, C. J. (ed) *Seeing is believing? Approaches to visual research* (pp. 105–121). Kidlington: Elsevier.

Felstead et al.'s chapter provides some very clear and accessible case studies in which they have incorporated photographs into their interview design.

Banaji, S. (2010). Seduced 'outsiders' versus sceptical 'insiders'?: Slumdog Millionaire through its re/viewers. *Participations: Journal of Audience and Reception Studies*, 7(1), 1–24.

Banaji's paper clearly illustrates approaches to interviewing about cinema and provides an excellent case study for incorporating non-verbal materials into interview practice.

Rose, G. (2016). *Visual methodologies: An introduction to researching with visual materials* (3rd ed). London: SAGE.

An essential book for all types of visual research. Rose covers a variety of research contexts, many of them involving interviewing, that involve using static and moving images.

REFERENCES

Banaji, S. (2010). Seduced 'outsiders' versus sceptical 'insiders'?: Slumdog Millionaire through its re/viewers. *Participations: Journal of Audience and Reception Studies*, 7(1), 1–24.

Boden, Z., Larkin, M. and Iyer, M. (2019). Picturing ourselves in the world: Drawings, interpretative phenomenological analysis and the relational mapping interview. *Qualitative Research in Psychology*, 16(2), 218–236,

Clark-Ibáñez, M. (2004). Framing the social world with photo-elicitation interviews. *American Behavioral Scientist*, 47(12), 1507–1527.

Dean, J. (2015). Drawing what homelessness looks like: Using creative visual methods as a tool of critical pedagogy. *Sociological Research Online, 20*(1), 1–16.

Dunne, J. H., and Pimlott-Wilson, H. (2021). Moodboards and LEGO: Principles and Practice in Social Research. In von Benzon, N., Holton, M., Wilkinson, C., and Wilkinson, S. (Eds) *Creative methods for human geographers*. London: SAGE.

Eldén, S. (2013). Inviting the messy: Drawing methods and 'children's voices'. *Childhood, 20*(1), 66–81.

Evans, A. B. (2008). Enlivening the archive: Glimpsing embodied consumption practices in probate inventories of household possessions. *Historical Geography, 36*(1), 40–72.

Glegg, S. M. (2019). Facilitating interviews in qualitative research with visual tools: A typology. *Qualitative Health Research, 29*(2), 301–310.

Felstead, A., Jewson, N., and Walters, S. (2004). Images, interviews and interpretations: Making connections in visual research. In Pole, C. J. (Ed) *Seeing is believing? Approaches to visual research* (pp. 105–121). Kidlington: Elsevier.

Folkestad, H. (2000). Getting the picture: Photo-assisted conversations as interviews. *Scandinavian Journal of Disability Research, 2*(2), 3–21.

Gallagher, M. (2015). Sounding ruins: reflections on the production of an 'audio drift'. *Cultural Geographies, 22*(3), 467–485.

Hall, T., Lashua, B., and Coffey, A. (2008). Sound and the everyday in qualitative research. *Qualitative Inquiry, 14*(6), 1019–1040.

Harrison, K., and Ogden, C. A. (2019). 'Grandma never knit like this': Reclaiming older women's knitting practices from discourses of new craft in Britain. *Leisure Studies, 38*(4), 453–467,

Henwood, K., Shirani, F., and Groves, C. (2018). Using photographs in interviews: When we lack the words to say what practice means. *The SAGE handbook of qualitative data collection* (pp. 599–613). London: SAGE.

Kelly, B. L. (2015). Using audio documentary to engage young people experiencing homelessness in strengths-based group work. *Social Work with Groups, 38*(1), 68–86.

Merriman, P. (2005). 'Respect the life of the countryside': The Country code, government and the conduct of visitors to the countryside in post-war England and Wales. *Transactions of the Institute of British Geographers, 30*(3), 336–350.

Peters, K. (2011). Negotiating the 'place' and 'placement' of banal tourist souvenirs in the home. *Tourism Geographies, 13*(2), 234–256.

Rose, G. (2004). 'Everyone's cuddled up and it just looks really nice': An emotional geography of some mums and their family photos. *Social and Cultural Geography, 5*(4), 549–563.

Rose, G. (2016). *Visual methodologies: An introduction to researching with visual materials*. London: SAGE.

Smith, T. A., and Dunkley, R. A. (2021). Video ethnography. In von Benzon, N., Holton, M., Wilkinson, C. and Wilkinson, S. (Eds) *Creative methods for human geographers*, (pp. 297–308). London: SAGE.

Strack, R. W., Magill, C., and McDonagh, K. (2004). Engaging youth through photovoice. *Health Promotion Practice*, *5*(1), 49–58.

von Benzon, N., Holton, M., Wilkinson, S. and Wilkinson, C. (2021). *Creative methods for human geographers*. London: SAGE.

Walsh, K. (2006). 'Dad says I'm tied to a shooting star!': Grounding (research on) British expatriate belonging. *Area*, *38*(3), 268–278.

Wang, C. C. (1999). Photovoice: A participatory action research strategy applied to women's health. *Journal of Women's Health*, *8*(2), 185–192.

Whittaker, G. R., and Peters, K. (2021a). Research music: listening and composing. In, von Benzon, N., Holton, M., Wilkinson, C., and Wilkinson, S. (Eds)s *Creative methods for human geographers* (pp. 217–228). London: SAGE.

Whittaker, G. R., and Peters, K. (2021b). Research with sound: An audio guide. In von Benzon, N., Holton, M., Wilkinson, C. and Wilkinson, S. (Eds) *Creative methods for human geographers* (pp. 129–140). London: SAGE.

Woodward, S. (2016). Object interviews, material imaginings and 'unsettling' methods: Interdisciplinary approaches to understanding materials and material culture. *Qualitative Research*, *16*(4), 359–374.

8

DIGITAL INTERVIEWING

Chapter objectives

This chapter examines the role of digital technology in enhancing and supplementing in-person interviewing. By reading this chapter, you should

- gain an understanding of what digital interviewing is used for,
- identify the challenges and opportunities of designing and implementing digital interviewing,
- examine a variety of different digital interviewing approaches, and
- understand the ethical, moral and risk challenges associated with digital interview practice.

INTRODUCTION

Throughout this book, I have privileged the interactivity of face-to-face interview encounters. There is a lot that can be drawn from an in-person interview, from the conversation itself to the non-verbal cues and environmental relations that can only be captured fully when face-to-face. Yet, it is impossible not to acknowledge the increasing role of technology in interview practice and the important rise in popularity of, what we can refer to as, 'digital interview practice' (Salmons, 2012, 2014; Shuy, 2003; James and Busher, 2009). Digital technology is present in every part of our everyday lives, meaning we do not just live 'with' technology, but we live our lives 'through' it. It is therefore somewhat obvious that digital technology would make its way into interview research design and practice, and in this

DOI: 10.4324/9781003292784-10

chapter, I will examine the ways in which technology can substitute in-person interaction, under what circumstances digital interviewing is appropriate and the digital methods that can enhance the interview encounter itself. Digital technology creates opportunities for interviews to be conducted flexibly over time and across much greater distances, so in this chapter, I will explore the pervasiveness of digital technology in contemporary interview design and practice, as well as outlining some of the opportunities and challenges that come from 'doing' interviews digitally.

In this chapter, I introduce you to the basics of digital interview practice. First, I outline what constitutes a digital interview, why digital interviewing practice is necessary and how it differs from in-person interviewing. Next, I list some of the key digital interviewing techniques. Third, I provide some advice on how to conduct digital interviews, including troubleshooting guidance. Finally, I propose some new ethical challenges that have arisen from digital interviewing and provide guidance for mitigating these.

WHAT IS DIGITAL INTERVIEWING?

Since the inception of the internet in the 1990s, interview practitioners have dabbled with incorporating digital technologies into interview design. This involved early versions of chatrooms, such as MUDs (Multi-User Domains), that provided interviewers the opportunity to engage with potential interview participants remotely (James and Busher, 2009). Due to the primitive nature of the technology, MUDs operated in much the same way as conventional discussions, in that the interviewer and interview participant needed to engage in a real-time conversation that was conducted in written form but in a virtual environment. These new approaches to interview practice were important in providing opportunities to recruit and engage with participants from vulnerable populations or from interest groups that might otherwise be difficult to find.

Yet, the nature of wired technology and the limited access many people had to personal computers meant that digital interviewing was considered niche and for the privileged few. It was with the advent of email and the subsequent invention of social media and the smartphone in the 2000s that technology became more available and accessible to global populations, and digital interviewing practices diversified and became more commonplace (Holton, 2021).

I outline a variety of digital techniques in this chapter, including email interviewing, video and audio-call interviewing, as well as interviewing that uses social media and instant messaging services. Hence, there now exist a far greater array of digital approaches that interviewers can build into their research design that either augment or replace altogether the conventional face-to-face encounter.

That said, while digital interviewing was gaining popularity in the 2000s and 2010s, it was not until the outbreak of the COVID-19 pandemic in 2020 that digital approaches became normalised in interview research (see Melis Cin et al.'s (2023) research in South Africa for a critique of the perception that online platforms homogenise interview practice). While national lockdowns and social distancing guidelines meant that almost all of everyday life shifted into the digital world during the pandemic – from Zoom quizzes with family members, increased online retailing, to home working and studying for many – the challenges associated with COVID-19 forced an enduring turn in how we live our lives digitally in a post-pandemic world. Academic research did not escape these changes, and qualitative interviewing practitioners that had relied for so long, and so deeply, on the in-person interactivity of doing face-to-face encounters had to adapt quickly to working remotely and with new technologies they potentially had little or no knowledge of ('t Hart, 2023). To say that digital interviewing – and we can confidently refer almost explicitly here to interviews using digital conferencing software (e.g., Zoom, Microsoft Teams, Skype) as being the primary digital approach being used – has taken off since the pandemic is an understatement, and digital interviewing is now likely on par with face-to-face interviewing as the common practice that most interviewers now use.

So, with digital technology now being embedded in contemporary interview practice, it is important that we consider the basics of these approaches to explore where the similarities and differences lay between face-to-face and digital interviewing and, more importantly, the new challenges that might arise from engaging with digital technologies.

WHY USE DIGITAL INTERVIEWING?

At this point, it is worth briefly exploring why digital approaches have been accepted into interview design. Adopting a digital approach should not be the result of a simple coin toss, and there are unique opportunities and challenges that come from using digital techniques. Digital approaches to interviewing have many benefits. They can be good for enabling long-distance participation, meaning interviews can be conducted with populations outside of the geographical constraints of the interviewer's capacities (Thunberg and Arnell, 2022). For example, you might be targeting a particular interest group that may be regionally, nationally or internationally spread, and a digital interviewing approach would therefore make it possible to capture the experiences of participants otherwise impossible to reach. This can make digital interviewing cost-effective, time efficient and potentially more sustainable by focussing the interview into a specific timeslot, thus reducing the time and cost of travel to and from an encounter. There may well be safety benefits too from digital interviewing in terms of avoiding interviews being conducted in unsafe or threatening physical locations.

Yet, digital interviewing can come with several pitfalls and issues (see Deakin and Wakefield, 2014; O'Connor and Madge, 2017). All digital approaches will require some level of technical competence with the technology being used. Providing simple instructions to participants on how to set up and engage with software (e.g., video conferencing platforms or mobile phone apps) or hardware (e.g., video cameras, computers or microphones) will help alleviate stress during an interview encounter and enable both you and your participants to concentrate on the conversation itself. Alongside technical competencies, technical difficulties are a primary issue for digital interviewing. Poor internet connection might result in video calls buffering or dropping out, while computer security settings might disable cameras or microphones. Testing equipment and having a troubleshooting guide should help fix problems if they occur. Technical difficulties can also hamper data quality, meaning parts of audio recordings might be lost or hard to hear, or video footage could become grainy or pixelated. If you are concerned that this might be happening, then discuss with your participant whether or not to pause the interview to check the quality of the data being produced. If issues arise, then there may be a chance to fix them and carry on without wasting the entire encounter. Finally, digital interviewing brings to the fore new ethical and security challenges. I will discuss these in detail later in this chapter, but it is important not to think of digital interviews being simply online versions of an in-person encounter. Conducting interviews over the internet will require you to think about a variety of challenges, such as data security and propriety, confidentiality and anonymity and recording and sharing, that do not have the same characteristics in person.

TYPES OF DIGITAL INTERVIEW

As with all interviewing styles, there is no single approach to digital interview practice, and the range of digital interviewing techniques can be broadly divided into two categories – synchronous interviewing (interviews conducted in real time) and asynchronous interviewing (interviews conducted in non-real time; O'Connor and Madge, 2017; James and Busher, 2009). Both approaches have their relative benefits and disadvantages, and in this section, I will outline some of the primary techniques that are often used in interview practice.

Synchronous interviews

Synchronous interview approaches most closely resemble the conventional face-to-face interview in that they are conducted in real time and in the presence of the interviewer and interview participant. This means that synchronous interviewing retains some of the interactivity of an in-person encounter, allowing for spontaneity and a more natural conversation than the asynchronous approaches outlined in the next section. The two common forms of synchronous interviewing I outline next involve audio (telephone) and video (video conferencing software) technologies, although other forms of synchronous technology exist, such as internet chatrooms that are, perhaps, less commonly used.

Telephone interviewing

Telephone interviewing has its roots in telephone survey interviewing – a technique that employs the telephone as a conduit through which to survey large samples of participants (Bourque and Fielder, 2003). In the context of this book, I discuss telephone interviewing in the same long-form, discursive manner as a conventional interview rather than as a method to administer short, quantitative surveys. Hence, as a technique, telephone interviewing can be a valuable tool in that it is cheaper than conducting face-to-face interviews, can be more flexible and convenient for interview participants and may well cover a wider geographical spread of participants than is possible to reach in-person (Novick, 2008). There may well be instances whereby your participant group is difficult to access physically, and/or they may not have sufficient access to or knowledge of the internet to conduct a video call. The downside is that the telephone is quite a disembodied device, meaning it is difficult to discern tone and intonation, and impossible to read anything into the body language of your participants. The telephone has also largely been taken over by digital technology as the most popular everyday form of communication, meaning some participants might not have access to, or feel confident using, a telephone.

Video interviewing

Video technologies have become increasingly popular as an interviewing tool (see Hine, 2004, 2008; James and Busher, 2006; O'Connor and Madge, 2017), and there are a variety of free, easy-to-use platforms available (see Oliffe et al.'s (2021) Canadian and Australian research on interviewing with Zoom). Video interviews carry many of the same benefits as telephone interviews in that they are generally cheap to administer, use readily available technology and collapse distance and time in ways that in-person interviews cannot. The added advantage of video technologies is the ability for both interviewers and interview participants to see each other, making it possible to pick up visual and non-verbal cues that telephone interviews cannot. Video interviews carry disadvantages, though, and even with the significant push towards digital communication since the pandemic, it is unwise to presume all participants can, and will want to, engage with video calling. Moreover, video calls come with privacy issues, and some participants may not have access to private spaces away from family or work colleagues to conduct an interview (Adams-Hutcheson and Longhurst, 2017; Deakin and Wakefield, 2014).

Asynchronous interviews

In contrast to the interactivity of synchronous approaches, asynchronous interviewing, that is interviews conducted in non-real time, are, perhaps, more flexible in nature and work in completely different ways to a conventional conversation. Asynchronous interviewing was extremely popular in the 2000s and 2010s as people became more accustomed to communicating regularly using email. This approach can be viewed as a more inclusive form of interview practice in that interview questions can be provided digitally, in written form, for a participant to reflect upon and answer in their own time without the pressure of being in the presence of an interviewer and voice recorder (O'Connor and Madge, 2017). Asynchronous interviews can also be iterative in that the questions and responses can be passed back and forth between interviewer and interview participant, with additional questions and responses posed or, if conducted via social media or instant message, taking the form of a digital conversation that can be added to over a longer period of time.

Email interviews

As outlined earlier, email interviewing has been one of the primary applications of asynchronous digital interviewing over the years, and the non-verbal and iterative nature of email has proved a popular method of research, particularly for those who prefer to communicate in written text (Dahlin, 2021). In practice, they can be administered all at once, with participants being emailed a list of questions from which to respond or, discursively, with a single, open-ended question being emailed and responded to, and the responses forming the basis of further emailed questions. Email interviews are therefore valuable in that they provide interview participants the space to own their own narrative without feeling pressurised into giving instant responses, but they also come with pitfalls due to their time constraints that may result in interview participants delivering superficial responses or becoming bored and dropping out (Burns, 2010).

Social media and instant messaging

Social media and instant messaging have become the natural successor to the email interview due to their ubiquity in everyday life and the range of potentially different platforms that provide opportunities to capture different types of interview data (e.g., video, photographs, GPS waypoints). Like email, social media has the capacity to develop an interview conversation over a longer time period, with the added benefit of being housed on participants' mobile technologies. This can make it easier for participants to engage with the process rather than having to interface with technology that they do not regularly use (Pearce et al., 2014). There can, however, be proprietary issues with social media interviews in terms of who owns the data that is produced, meaning confidentiality needs to be considered carefully when using existing platforms.

HOW TO 'DO' DIGITAL INTERVIEWS

As with face-to-face interviewing, there are many nuanced approaches to working with digital interview practices, meaning you are likely to need to consider the intricacies of your project and carefully weigh up if and how digital technologies might be appropriate for your interview design. That said, there are fundamental basics to each of the interview types I outlined earlier, and I will explain these here.

Telephone interviews

As I mentioned earlier, the telephone interview approach that I refer to here differs from the interview surveying techniques used by telemarketers to gain quantitative data. Telephone interviewing follows a similar ethos to a face-to-face encounter or to the video techniques I will discuss next. You would likely use a telephone interview approach if your participant was unable, or unwilling, to meet in person, and in some circumstances, telephone interviews are used as an alternative research method to include such participants in the research (Novick, 2008; Block and Erskine, 2012). There are clear compromises in adopting telephone interviewing (e.g., relying on just the voice), but telephone interviews can be a useful way of collecting data quickly and conveniently.

There are some obvious safety and security issues that need to be addressed before a telephone interview commences. First, it is important to check that participants are in a safe environment and that they are not undertaking any other activity that might otherwise require their full attention. For example, telephone interviews must not be conducted while the participant is driving or walking, or if they are operating machinery. This sounds obvious, but you may not be aware of this until the interview has started and sometimes interview participants might feel tempted or compelled to 'squeeze you in' among other daily tasks. Moreover, it is important to know where the participant is located and whether the space within which they are conversing with you is private and secure. If, for example, an interview participant is in a public space, they may feel exposed and vulnerable if asked to respond to a sensitive question. This might not be such a problem in the discursive context of a conventional face-to-face interview, but participants might feel exposed when alone and on the telephone.

The majority of contemporary telephone interviews are likely to be conducted on mobile phones, so ensuring that both parties have a good mobile signal is going to be important. As noted earlier, ensuring participants are stationary and comfortable should help retain a phone signal. One of the difficulties associated with telephone interviews is recording. This can be as simple as putting a phone on loudspeaker and placing a voice recorder next to the handset, but more appropriate technology now exists, such as telephone pick-up devices that link to an existing voice recorder and capture both the voices of the interviewer and interview participant. As with all interviews, it is good practice to check this technology ahead of the interviews to make sure you are familiar with the controls and that the audio quality is sufficient for transcribing.

Video interviews

In many ways, video conferencing software has been championed as a panacea for many of the issues with traditional face-to-face interviews outlined in this book. Yet, it is important not to presume video interviews to be problem-free, and any interview research that includes a video component needs to be carefully weighed up beforehand. Deakin and Wakefield (2014) present three key considerations for designing video interviewing.

1. The logistics of the interview need to be understood. Video interviews can be convenient in terms of when they can take place and where, and this can be attractive to participants when recruiting. It will be important, though, to discuss the process with participants beforehand to ensure both parties have good internet access, are located in safe and private environments to take a call, and have agreed on the method of recording the interview (e.g., whether to have the video on or off during the call).

2. Ethics needs to be carefully considered in video interviewing in terms of privacy, confidentiality and identity. It may well be that the ethical considerations that are deemed acceptable in face-to-face contexts do not easily map onto digital approaches that use video in terms of the method of conducting and recording the interview. It is, therefore, important to develop a framework that relates to the platform that you are using and to share this with interview participants ahead of the interview being conducted.

3. Rapport is often handled very differently online than it is face-to-face. The video screen can act as a bridge to participants that enables the nuances of emotions and body language to come through that would not be possible over the telephone or using email. Yet, screens can also act as an awkward barrier between the interviewer and the interview participant (Adams-Hutcheson and Longhurst, 2017). Hence, as noted earlier, do not presume that participants will be comfortable using the video function and prepare for some participants to want to turn off the video feed or hide their own video image during the call.

In addition, participants may need training prior to using the software or be given access to new software that they need to download onto their computers. This can raise issues around privacy, particularly if participants are talking to you in public spaces or among others in their homes who are not involved in the research. Moreover, there may be technical considerations, like internet bandwidth, that interrupt the call or even terminate it completely. Hence, being upfront about what you want this type of technology to do, and what to do if things go wrong is important in ensuring your participants feel as comfortable as possible.

The other element that is often unique to video interviewing is negotiating what to record during an interview. I noted earlier and in Box 8.1 that it would be wrong to assume that your participants will be automatically happy to have their interview video-recorded, and it is vitally important that consent is gained prior to recording taking place. The most crucial thing here, though, is to consider whether a video recording will add value to the subsequent transcript and analysis. I discuss elsewhere in Chapter 4, and later in Chapter 9, how non-verbal cues can help add context and meaning to interview responses, so therefore, having a video record of an encounter may well allow you to examine these cues in situ without relying on incomplete notes or memory. Moreover, you may have incorporated other elements into the research encounter, such as using a screen-sharing function to display images, videos or other online materials (Maulana, 2023). Capturing the video might then mean that you can follow the sequence of events alongside the transcribed material to ensure accuracy throughout. Yet, while video conferencing software may yield benefits in triangulating the data produced and perhaps building a stronger sense of rigour into interview practice, video recording can be an even more anxiety-ridden process for interview participants than the voice recorder and can, in many ways, disembody the research process by distancing the interviewer and interview participant (Adams-Hutcheson and Longhurst, 2017). It is important therefore that the recording process and data handling procedures are explained clearly to participants.

Box 8.1   Netiquette using video conferencing software

There can often be a number of assumptions when using video conferencing software for interviews that everyone is familiar with the technology and the practice of talking online. This is certainly not the case, and technologies like Zoom, Microsoft Teams and Skype can be very off-putting for some potential participants if they are unsure how to use the interface and they may become uncomfortable with the activity itself. Below are some useful tips from the software developer Zoom that can help troubleshoot potential problems:

1. Practice using the technology and acquaint yourself with the functions you need for the interview. This will increase your confidence when using the software and prevent issues from arising that might affect the quality of the recording.
2. Provide your participants with briefing notes or instructions ahead of the interview so that they understand how to access and use the technology. Keep these very simple, and remember that some software might not look exactly the same across all computer platforms. Check with your participants that they are comfortable and able to work the software before recording.
3. Practice speaking to the camera rather than the person on the screen. This can help make the encounter warmer and more personal.

4. Check that your participants are happy for you to record the inter-view and are willing and able to have their cameras turned on. Do not assume that your participants will automatically be okay with this; there may be a variety of reasons why a participant can-not have their camera turned on, such as low bandwidth or a lack of privacy. Encourage your participants to blur their backgrounds using the built-in functions of the software.

5. Be prompt, and do not leave your participants waiting to join a meeting. Moreover, avoid starting the interview straight away. Take time to settle your participants into the encounter, particu-larly as they may be nervous speaking online.

6. As with face-to-face interviews, state when you start and end the recording so that participants are clear on what is and is not being included.

7. Remember your own behaviour and body language. Avoid yawn-ing, fidgeting or checking your phone. Stay attentive as you would in an in-person encounter.

(See Zoom's (2019) "Video Meeting Etiquette: 7 Tips to Ensure a Great Attendee Experience", https://blog.zoom.us/video-meeting-etiquette-tips/.)

Email interviews

As I mentioned earlier in the chapter, email interviews can be an excellent way of interviewing individuals who may struggle with face-to-face conversation and might prefer articulating their thoughts in written form and in their own time.

The 'all-in-one' method is an approach similar to online surveying, whereby participants are provided with a list of questions to respond to. This is potentially easier than setting up approaches that require questions and responses to be sent back and forth over longer periods of time. Yet, providing all the questions at once can be overwhelm-ing for participants, particularly if they feel time-pressured to answer them in one block (Dahlin, 2021; James and Busher, 2006). This can have the negative effect of producing shorter, diluted responses that can be difficult to derive meaningful analysis from. In these sorts of scenarios, it might be better to consider an alternative approach, such as an online survey, that could be sent to larger sample group.

Providing participants with single questions, or a brief collection of questions relating to a specific theme, will more likely generate the necessary depth required for analysis. You might, for example, send all of your participants the same introductory question(s) but then adapt subsequent questions according to the type of responses you receive back. This would mimic, more closely, the conversational qualities of a conventional face-to-face interview and allow participants to retain control over what knowledge they discuss and how they present it to you.

The following is a checklist of things to think about when using email interviewing:

1. It is important for participants to understand what the research is about and what is being expected of them, so provide them with some written context for the purpose of the interview. This may be written in your invitation letter or sent in a preparatory email.
2. Supply very clear instructions and guidelines ahead of the commencement of the interview. These might include encouraging participants to provide responses of a particular length.
3. Set a clear deadline for responses to be returned. With all the best intentions in the world, it is easy to let emails slip down the list of priorities. Be flexible, though, and make adjustments if participants think they are unable to return in your timeframe.
4. As with conventional interviews, always ask open-ended questions. I recommended the 'tell me about…' question in Chapter 3, as adopting this sort of practice will limit the chance of responses coming back with simple 'yes' or 'no' answers.
5. Pose questions using simple language. Avoid jargon or acronyms in case they are misunderstood or unclear. Participants are unlikely to query email interview questions if they are already short on time, and poorly phrased or unclear questions might result in a participant dropping out of the process.
6. Provide space for further responses or additional information. There may be something that participants would like to add that you have not considered, and this might help stimulate further questions down the line.

Social media and instant messaging

Social media and instant messaging interviews follow a similar ethos to email interviewing. Where they are unique is that their 'mobile' characteristics often lend themselves to shorter, chattier responses, more akin to regular conversation (Wilkinson, 2016; Halliwell and Wilkinson, 2021). This means that social media interviews might generate longer-form conversations that comprise lots of shorter responses. Participants are also likely to be more familiar with interacting with social media chat (they might even be put off by the perceived formality of email), meaning the tone of conversation, types of responses and methods for conveying information (e.g., using emojis, 'text-speak', acronyms) can mimic the identities of the participants you are working with. Moreover, the ease at which people engage with social media might mean that they operate both asynchronously, as and when participants log into their apps, or synchronously if both participant and interviewer and logged in at the same time and exchange messages in real time (O'Connor and Madge, 2017).

I will outline some of the ethical and moral dilemmas associated with digital interviewing next, but one of the primary things to consider with social media, in particular, is that you are likely to be operating in a fairly public environment. Platforms like Facebook and Instagram have dedicated private messaging features, and it would be advisable to direct the conversations into these spaces rather than conducting them in public. There are privacy issues involved here in terms of making connections and considering whether to retain them after the interview process is complete. You may have to formally 'add friends' or 'followers' to your own private network, and your participants will have to do the same. There are moral risks here in inviting strangers into your own private domain, so weigh up whether setting up a dedicated account for the project might be appropriate. Moreover, also consider how you will harvest the data provided in the chat. WhatsApp, for example, has a feature that allows chat to be exported into Microsoft Word and Excel files. It is important to check if there is a similar approach for your own platform ahead of conducting any research in case it becomes difficult or impossible to extract the information into a form that be analysed.

Box 8.2 Social media interview checklist

In this box, I have adapted some guidance provided by LinkedIn (2023) on how to design and implement effective social media interviews.

1. Choose the right platform for your research. Look at the features of the social media platforms available to you to help you decide on which platform best suits the research you are conducting. Consider the following questions. Is the platform appropriate for the type of interview participants you are targeting (e.g., think about the size and composition of the type of audience that uses the platform)? What method of communication does the platform provide (e.g., public forum posts, private messaging etc.)? Is the platform easy and/or intuitive for participants to use?
2. Explore the privacy settings within your chosen platform. Do they offer sufficient privacy for your participants to feel comfortable sharing information without risk of harm?
3. Prepare your questions. Make sure the questions you post are easy to follow and respond to. The trick here will be to write these clearly and concisely. As with in-person interviews (see Chapter 3), avoid leading questions or double questions, but do keep your questions open so as to encourage longer responses.
4. Set out clear instructions for engagement. This will need to include things like timeframes for responding, the types and preferred length of responses and netiquette (see Box 8.1).
5. Stick to your rules! If you tell participants that you will respond within 24 hours, make sure you do. If there are to be periods when you cannot respond, then let participants know.
6. Engage your interview participants. Be polite, friendly and professional in your conduct, and show interest in the responses they are posting. There is a significant time commitment to running social media interviews so make sure your interview participants understand your appreciation.
7. Have a method for recording your data. The platform may have a feature that allows you to download messages, but you might have to collect screenshots or transcribe messages into a word-processed document or spreadsheet. However you do this, be accurate, complete and consistent.

'DOING' DIGITAL INTERVIEWS ETHICALLY

Digital forms of interview practice inevitably bring with them allied challenges in how we consider ethics and risk; implications concerning anonymity, confidentially and safety; and what types of knowledge are being produced. How, for example, might interviewing digitally influence our relationships with participants? What ethical challenges are unique to digital interview techniques?

Fundamentally, digital interview practices concern new, and often untested, approaches to considering ethics and risk. Interviewing with digital technologies involves researching in multiple environments (the technological apparatus, the virtual platform being used and the geographical space the interview is conducted in), and each will likely present particular challenges for conducting research (Salmons, 2016). Kinsley (2021) cautions that digital research, including interviewing, has challenges that are unlikely to be experienced in conventional 'real-world' contexts. These include the negotiation of public and private information, the management of informed consent and the establishment of confidentiality and anonymity.

Social media is a classic example of the slippery nature of the public and the private in terms of public posts and private messages (see Halliwell, 2023). When drawing social media into interview practice, it is important to be clear how participants will negotiate the platform(s) being used and where and when the research commences and ends. This needs to be carefully and explicitly stated in the informed consent provided to participants to ensure they understand precisely what is expected of them, what behaviours and acceptable and unacceptable online and what to do if they feel uncomfortable or unsafe. Halliwell and Wilkinson (2021) discuss this in the context of 'flaming' (the practice of being abusive to others online) and provide supportive tips for negotiating behaviour when working with participants online.

Yet, while we as researchers can build diligence into our own research practice, we may not necessarily know whether the technologies we want to use follow similar, or indeed any, ethical considerations (Salmons, 2014). Using third-party smartphone apps is an example of instances where institutional ethical approval can be undermined. How, for example, do we know if the content our participants upload into a third-party app is private and anonymous?

Might data, including personal details, be harvested through the app and used for other commercial purposes? Who 'owns' the photographs, video footage or text that might be uploaded into the app? How can we be certain that this content is not going to end up in marketing material for the company that produced the app? This all seems rather catastrophic, and I certainly do not intend to put you off using apps, but it is crucial that researchers do not simply presume technology to be value-free or unproblematic.

One of the key concerns that has emerged from interviewing in online spaces is the moralities associated with using participants' personal technologies for research purposes. As stated earlier, the ubiquity of mobile and smartphone technologies often leads researchers to consider them as value-free and unproblematic devices (Beddall-Hill et al., 2011), whereas in reality, using personal technologies in interviews repurposes them from being private devices into research tools (Holton, 2021). In my own research, I have utilised a variety of free smartphone apps to help enliven interview practices, encouraging participants to track pedestrian walking using the GPS embedded in their phones (Holton and Riley, 2014), engaging with walking tour apps (Holton and Harmer, 2019; Holton, 2019) and producing digital scrapbooks (Holton, 2023). Central to my use of apps is a sense that I, as the researcher, have privileged access to my participants' technologies, be it in terms of accessing the content produced using, and stored within, mobile phones, as well as having control over how participants interact with their phones during the research period.

SUMMARY POINTS

- Contemporary interview design and practice have been greatly influenced by digital technologies.
- Digital technologies provide opportunities for interviews to become more convenient, cost-saving and sustainable.
- Technology has changed the ways in which we conduct interviewing, such as synchronous approaches that use video conferencing software and mobile phones that perhaps free up the interview encounter, and the use of asynchronous approaches, like email and social media interviewing.
- Digital interviewing approaches have implications for ethics, risk and positionality in terms of what types of new considerations

need to be incorporated into interview design and the ways in which online identities are mediated.

WHAT TO DO NEXT

- Think about what you want to achieve from a schedule of digital interviews. Remember that digital interviewing can mimic in-person interviewing but there are crucial differences in the various approaches outlined in this chapter.
- Consider whether both you and your participant group have access to relevant technologies and have the necessary competencies to operate them.
- Prepare a set of instructions for your participants to help guide them through the process and have some troubleshooting guidance to help mitigate potential problems.

Suggested further reading

Kinsley, S. (2021) Virtual spaces and social media. In Wilson, H. F., and Darling, J. (eds) *Research ethics for human geographers* (pp. 269–279). London: SAGE.

    Kinsley's chapter takes an inquisitive look at the ethical considerations behind online and virtual research methods. The chapter is extremely useful for researcher considering incorporating social media or third-party apps into their interview design.

O'Connor, H., and Madge, C. (2017). Online interviewing. In Fielding, N. G., Lee, R. M., and Blank, G. (eds) *The SAGE handbook of online research methods* (2nd ed., pp. 416–434). London: SAGE.

    This chapter details the practical considerations for choosing and implementing a range of different digital interviewing techniques. The authors provide particularly good advice on the potential and limitations of adopting digital interviewing practices.

Thunberg, S., and Arnell, L. (2022). Pioneering the use of technologies in qualitative research–A research review of the use of digital interviews. *International Journal of Social Research Methodology*, 25(6), 757–768.

    Thunberg and Arnell present clear and accessible guidance for the possibilities and limitations of online interviewing methods, focussing on quality, convenience and sensitivity.

REFERENCES

Adams-Hutcheson, G., and Longhurst, R. (2017). 'At least in person there would have been a cup of tea': Interviewing via Skype. *Area, 49*(2), 148–155.

Beddall-Hill, N., Jabbar, A., and Al Shehri, S. (2011). Social mobile devices as tools for qualitative research in education: iPhones and iPads in ethnography, interviewing, and design-based research. *Journal of the Research Center for Educational Technology, 7*(1), 67–90.

Block, E. S., and Erskine, L. (2012). Interviewing by telephone: specific considerations, opportunities, and challenges. *International Journal of Qualitative Methods, 11*(4), 428–445.

Bourque, L. B., and Fielder, E. P. (2003). *How to conduct telephone surveys.* London: SAGE.

Burns, E. (2010). Developing email interview practices in qualitative research. *Sociological Research Online, 15*(4), 24–35.

Dahlin, E. (2021). Email interviews: a guide to research design and implementation. *International Journal of Qualitative Methods, 20*, 1–10.

Deakin, H, and Wakefield, K. (2014) Skype interviewing: Reflections of two PhD researchers. *Qualitative Research, 14*(5), 603–616.

Halliwell, J. (2023). 'Are you sure you're not gay?': Straight and bisexual male experiences of Eurovision Song Contest fandom. *Social & Cultural Geography, 24*(6), 1024–1041.

Halliwell, J., and Wilkinson, S. (2021). Mobile phones, text messaging and social media. In, von Benzon, N., Holton, M., Wilkinson, C. and Wilkinson, S. (Eds) *Creative methods for human geographers* (pp. 259–272). London: SAGE.

't Hart, D. (2023). COVID times make 'deep listening' explicit: Changing the space between interviewer and participant. *Qualitative Research, 23*(2), 306–322.

Hine, C. (2004). Social research methods and the Internet: A thematic review. *Sociological Research Online, 9*(2). https://openresearch.surrey.ac.uk/esploro/outputs/99513010302346?institution=44SUR_INST&skipUsageReporting=true&recordUsage=false. accessed 20th November, 2023

Hine, C. (2008). The internet and research methods. In Gilbert, N. and Stoneman, P. (Eds) *Researching social life* (4th ed., pp. 304–320). London: SAGE.

Holton, M. (2019). Walking with technology: Understanding mobility-technology assemblages. *Mobilities, 14*(4), 435–451.

Holton, M. (2023). *Encountering coastal youth citizenship: Exploring young people's engagements with coastal environments in the UK. Final project report.* London: RGS-IBG.

Holton, M. and Harmer, N. (2019). 'You don't want to peer over people's shoulders, it feels too rude!' The moral geographies of using participants' personal smartphones in research. *Area, 51*(1), 134–141.

Holton, M., and Riley, M. (2014). Talking on the move: Place-based interviewing with undergraduate students. *Area*, *46*(1), 59–65.

Holton, M. (2021). Creating and repurposing apps. In von Benzon, N., Holton, M., Wilkinson, C., and Wilkinson, S. (Eds) *Creative methods for human geographers* (pp. 273–284). London: SAGE.

James, N. and Busher, H. (2006). Credibility, authenticity and voice: Dilemmas in online interviewing. *Qualitative Research*, *6*(3), 403–420.

James, N., and Busher, H. (2009). *Online interviewing*. London: SAGE.

Kinsley, S. (2021) Virtual spaces and social media. In Wilson, H. F., and Darling, J. (Eds) *Research ethics for human geographers* (pp. 269–279). London: SAGE.

LinkedIn. (2023). *How do you conduct social media interviews?* https://www.linkedin.com/advice/3/how-do-you-conduct-social-media-interviews-skills-social-media. accessed 15th November, 2023.

Maulana, M. I. (2023). Leveraging Zoom video-conferencing features in interview data generation during the Covid-19 pandemic. In Cahusac de Caux, B., Pretorius, L. and Macauley, L. (Eds) *Research and teaching in a pandemic world: The challenges of establishing academic identities during times of crisis* (pp. 391–407). Singapore: Springer Nature.

Melis Cin, F., Madge, C., Long, D., Breines, M., and Tapiwa Beatrice Dalu, M. (2023). Transnational online research: recognising multiple contexts in Skype-to-phone interviews. *Qualitative Research*, *23*(2), 252–271.

Novick, G. (2008). Is there a bias against telephone interviews in qualitative research? *Research in Nursing & Health*, *31*(4), 391–398.

O'Connor, H. and Madge, C. (2017). Online interviewing. In Fielding, N. G., Lee, R. M., and Blank, G. (Eds) *The SAGE handbook of online research methods* (2nd ed., pp. 416–434). London: SAGE.

Oliffe, J. L., Kelly, M. T., Gonzalez Montaner, G., and Yu Ko, W. F. (2021). Zoom interviews: Benefits and concessions. *International Journal of Qualitative Methods*, *20*, 1–8.

Pearce, G., Thøgersen-Ntoumani, C., and Duda, J. L. (2014). The development of synchronous text-based instant messaging as an online interviewing tool. *International Journal of Social Research Methodology*, *17*(6), 677–692.

Salmons, J. (Ed) (2012). *Cases in online interview research*. London: SAGE.

Salmons, J. (2014). *Qualitative online interviews*. London: SAGE.

Salmons, J. (2016). *Doing qualitative research online*. London: SAGE.

Shuy, R. W. (2003). In-person versus telephone interviewing. In Holstein, J. A. and Gubrium, J. F. (Eds) *Inside interviewing: New lenses, new concerns* (pp. 175–193). Thousand Oaks: SAGE.

Thunberg, S., and Arnell, L. (2022). Pioneering the use of technologies in qualitative research – A research review of the use of digital interviews. *International Journal of Social Research Methodology*, *25*(6), 757–768.

Wilkinson, S. (2016). Hold the phone! Culturally credible research 'with' young people. *Childre's Geographies*, *14*(2), 232–238.

PART III

HANDLING INTERVIEW DATA

9

WHAT HAPPENS NEXT?
On transcribing, coding and analysis

Chapter objectives

This chapter will guide you through the post-interview process of transcribing, coding and analysing your data. By reading the chapter, you should

- recognise the importance of accurate and consistent transcription in aiding the analysis process,
- gain knowledge of the different approaches to coding and how and when to apply these to interview analysis,
- understand how to transform coding into effective and meaningful analysis using a range of techniques, and
- appreciate the benefits and pitfalls of utilising data analysis software.

INTRODUCTION

Transcription, coding and analysis could warrant an entirely separate book, so this chapter aims to provide a broad overview of what to expect from each element of the post-interview process and signpost useful in-depth guides and resources that are available in print or online. Getting from the point of completing data collection to writing can seem like a daunting process in all forms of qualitative research, something, according to Silverman (2015: 110) that is akin to "exploring a new territory without an easy-to-read map". If your skills and experience sit with quantitative analysis – with ordered spreadsheets and structured statistical analysis techniques – then the

DOI: 10.4324/9781003292784-12

data produced through interviews may, at first, appear disordered, and the methods for interpreting them seem disparate, contrasting and occasionally conflicting. This lack of consistency is, in fact, intentional. Interview material, as with all qualitative data, is highly subjective, meaning the methods for interview analysis might need to be subtly tailored depending on the research design, the phenomenon under investigation, the types of research questions posed, and the methods used to collect data.

In this chapter, I introduce you to some of the key post-interview processes associated with transcribing, coding and analysing your interview data. First, I explore transcription and why this is such an important stage of any interview process. This involves data management techniques, handling confidentiality and anonymity, the planning and structuring of transcripts and how to represent your participants' 'voices' on the page. Next, I briefly outline what we mean by data analysis and what is involved in the analysis process. Third, I introduce coding approaches. Here, I observe the differences between 'a priori' (pre-planned coding) and 'grounded theory' (free-form coding) approaches, as well as how to utilise 'axial coding' to formulate patterns and examine contrasts and relationships in the data. Alongside this, I briefly outline three data analysis frameworks – content analysis, thematic analysis and narrative analysis – that can help structure the analysis process. Finally, I provide some commentary on the relative strengths and weaknesses of adopting qualitative data analysis software into the analysis process.

TRANSCRIBING INTERVIEWS

This section weighs up what transcribing means and how to approach it, as well as addressing the question of completeness. Transcription is often viewed as a 'behind-the-scenes' activity (Oliver et al., 2005; Hepburn and Bolden, 2017), when in actual fact, transcribing interviews is a fundamental part of the interview process in which the interviewer/transcriber has the opportunity to reflect upon and interrogate the information as it transfers from spoken to written word (McLellan et al., 2003). To help provide context for the transcribing process, I include an anonymised example of a transcript from my own research to visually illustrate how techniques such as using pseudonyms, anonymising identifying information and recognising non-verbal gestures might be incorporated into your own transcripts (see Figure 9.1).

The process of interview transcription is very often one of the most overlooked elements of qualitative research (Kvale, 2007; Stuckey, 2014). I have read countless research proposals that simply state 'transcribe interviews' in their research schedule without really considering quite how much time, effort and engagement is required in preparing, undertaking and managing verbal information into a presentable and readable format (Crang, 2005; Clark et al., 2019). "My fingers hurt" is therefore a very common complaint I receive each year from my undergraduate dissertation students as they diligently transcribe their interviews. Data transcription software can, of course, speed up the transcription process, and there are indeed many free or low-cost software options that can transcribe audio files (e.g., Otter.ai, Microsoft Word, Zoom etc.), albeit often requiring some editing on behalf of the transcriber. While technology can significantly speed up the transcription process, the action of typing out your own interviews can, in many ways, constitute the first layer of analysis as you re-familiarise yourself with each encounter and begin to establish important themes that cut across your dataset (Cope and Kurtz, 2016). Transcribing interviews is therefore an unashamedly time-consuming process. Hence, I encourage you to sit down, plug in some headphones, familiarise yourself with the 'pause' button on your voice recorder and start typing!

A word on managing and storing your data

In Chapter 5, I briefly introduced the importance of establishing a robust data management plan in order to store data safely and securely. I revisit this here to argue that data management should not be considered a formality in research design but should influence how data are handled throughout the lifetime of the project. Ensuring data management is carried out and maintained effectively and consistently will, therefore, pay dividends at the transcribing and analysis stage of the research process. Qualitative interviews, by dint of being verbal information, will inevitably produce vast quantities of data. For example, the average 1-hour-long interview is likely to generate a written transcript of approximately 10,000–12,000 words in length. This, multiplied across, say, 15 interviews, will produce

Sarah: Yes, but then I suppose it's a social thing. The hour in-between, their catching up with people. Whereas I think, if I do this now, I haven't got to do it later so, I'm not being rude, but I don't want to sit and talk [*laughs*]. Sometimes I do. I wouldn't be rude to anyone but I'd rather make better use of my time. And whilst [*University building*] is not brilliant for studying, it's not worth me going to the library for an hour and back again, so I don't do it. But I do tend to sit and read something. I've always got spare uni [*sic*] stuff that I could be working on so yeah, I'll work when I've got gaps. And sometimes things will get cancelled when you're in so I tend to make use of that time as well]

Mark: Is this quite a solitary task?

Sarah: Yes … I don't think everyone has got the same work ethic as mature students [*laughs*].

Mark: You said that you have made connections with some younger undergraduate colleagues. How did you establish that?

Sarah: Because I'm doing [*subject*], one of them is doing exactly the same as me and we're in a lot of the same things. They do call me mum [*laughs*] and they do ask for my work [*laughs*]. I do give it to them after, when it's been marked. I do think it's quite good, I do get quite good marks and they just want to learn from me. So that's where it started, they started to see what I got and they thought "oh, I think I'll be friends with you". I don't mind that. It's good to be shallow sometimes. I have no thoughts that these friendships will go on or go outside. I don't socialise with them. They have asked me a couple of times but they are just being polite. They don't really want to go out with me [*laughs*] and I don't really want to go out with them, but they're funny and they remind me of being nineteen so.

Mark: You're fitting in without really fitting in?

Sarah: Yeah, I don't really want to fit in, I have no real desire to be nineteen again.

Annotations pointing to the text:
- Verbatim transcription
- Wide margin and double spacing for annotation / coding
- Notes about dynamics of research encounter: laughter, body language etc.
- Indications of interruption, hesitation, pauses, repetition etc
- Anonymised information
- Pseudonym

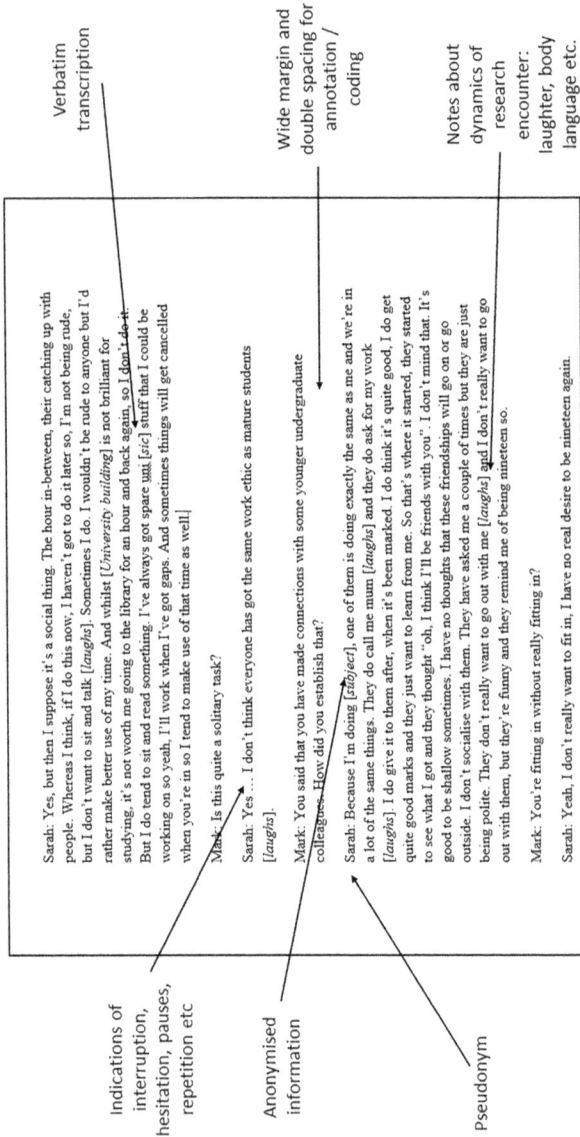

Figure 9.1 Structuring an interview

around 150,000–180,000 words of data. This is probably rather daunting right now, but having data this rich is essential in exploring qualitative research problems in depth.

An effective data management plan with therefore make navigating interview transcripts a much easier and rewarding process. Ensure your digital files are clearly labelled using a consistent scheme comprising the date and time of the interview, the anonymised 'name' of the participant and any other useful information, such as interview location or participant group. These data must also be securely backed up if stored digitally or locked away if printed out. Many institutions and organisations have ethical guidelines that usually have a very direct stance on how they want data management handled, so it is important to familiarise yourself with that guidance as you prepare your research.

Anonymity and confidentiality

A word of caution at this point is to reflect back on what terms of confidentiality and anonymity your participants agreed to when they signed a consent form prior to their interview (see Chapter 5). If you stated in your informed consent that your participants would be anonymous, then you must ensure you stick to this during the transcribing, analysis and presentation of data. There might be circumstances whereby an interview participant waives the right to be anonymised (e.g., the CEO of an organisation) or the research centres on a specific location or company. In these situations, the parameters of anonymity should have been agreed upon prior to the interviews being conducted.

Ensuring anonymity means that all personal information must remain unknown to anyone else but the researcher (Hall et al., 2021). Anonymisation is therefore a key consideration when transcribing interviews, particularly in terms of confirming the safety and confidentiality of your participants' responses as their data are worked up into analysis and finally represented in the final documentation. Thankfully, there are some typical conventions that you can consider adopting into your transcription practice before you begin.

Anonymisation can take on many forms, and this will usually need to be considered in relation to the specifics of your own research. Participants' names are usually the first things that need to be changed. This is sometimes referred to as 'de-identifying'

(Stuckey, 2014), whereby researchers provide their participants with a pseudonym – essentially a fake name – at the point of transcription that differs from their given name. This can sound straightforward, but it is important not to provide a pseudonym that turns out to be another participant's real name, so think carefully about how you choose these. Likewise, providing pseudonyms that are very similar to one another (e.g., Mark, Mary, Maya, Matt) can make analysis confusing to navigate. I have seen a variety of interesting, and at times quite elaborate, strategies used in past research, including re-naming participants after the characters from the TV show *Friends*, players in the 2015 Manchester United football team and the surnames of 20th-century British prime ministers. While this sounds complex in itself, the intention of this process, beyond confirming anonymity, is to ensure you are able to map easily and accurately through your data when analysing and still retain a sense of whom the information you are reading is attributed to once these data are presented in your final research output.[1]

Beyond pseudonyms for participants' names, you may also need to anonymise other identifying information within your interviews. If the location of your interview (i.e., a building, like a school or place of work, or a neighbourhood, town or city) could reveal the identities of your participants, then this may also need to be changed. This might involve simply redacting information in the transcript (e.g., "I have always looked upon my time working at [organisation] very fondly..."), but if the place features significantly in the research design, then perhaps providing a pseudonym might be advisable (e.g., "It's crazy to think I've lived in Middletown for such a long time..."). Moreover, consider carefully the implications for other pieces of information that might identify a participant – such as names of children, friends or pets; street names; job roles; or niche leisure activities. This may require devising a checklist and sorting back through your transcripts if new information reveals itself as you progress through the transcription process.

Structuring your transcripts

After anonymity, developing a clear and consistent strategy for typing up your transcripts should be the next priority. This sounds obvious, but as you are likely to be transcribing over a period of time, it can sometimes be easy to accidentally change the style, structure and format as you go. While this should not have an impact on a single transcript, once you begin to analyse across transcripts any inconsistencies can start to disrupt how the analysis is managed. Figure 9.1 outlines some useful tips for establishing consistency. This is not by any means the only way to format a transcript, and your teacher or advisor is likely to also have some excellent strategies that you may want to adopt (see Oliver et al. (2005) for some tips on transcription styles). What is important is that you find a style and stick to it. If, for example, you intend to print out physical copies of transcripts to annotate and code by hand, then consider using formatting tools like wide margins, double-line spacing, page numbers and including pseudonyms in page headers. Providing space in the document will offer you greater freedom in annotating the text once you begin to analyse (see Box 9.1).

You may also need to include non-verbal or environmental information that occurred during the interview in your transcript (Bailey, 2008). How, perhaps, did the research encounter itself influence the

---

Box 9.1   Piloting a transcript

As with the earlier points around anonymity and format, the style and detail of the transcription process should be guided by the intention of the data. If this is something that concerns you, perhaps try transcribing a section of the recording verbatim to produce a few pages of text and then review the results.

Here is a useful checklist of things to consider:

- Is the transcript structured logically? *Check to make sure you can distinguish between your questions and your participant's responses.*
- Have I anonymised my participant enough in the transcript? *Check for identifying information like names, places or organisations.*
- Can I 'hear' my participant's voice as I read through the transcript? *Perhaps try listening to the recording again while you read.*
- Does the transcript need to contain any non-verbal information? *Listen out for the way the participant responds to questions, the inflection of their voice or things like interruptions or pauses.*
- Does the transcript make sense to you once you read it back through? *Can you follow the conversation when you read it back?*

---

response(s) given by your participants? Might this affect how the data is understood once it is presented in the final output? Think about whether or not you need to include observations of the dynamics of the research encounter, such as laughter, sighing, crying, shouting or non-verbal cues, such as nervousness or hostile body language. If you are transcribing very soon after your interview, then annotating your transcript with these clarification details can sometimes add valuable context to the discussion. Stylistically, this information should be presented differently from the interview responses themselves. The following is an example whereby the supplementary information – in this case a visual and a non-verbal prompt – is placed in square brackets and italicised for emphasis.

> I mean obviously we've got that building opposite [*turns and points to across the road*]. This used to be a big leisure centre where we used to go swimming, stuff like that, or go for a coffee [*sighs*]. There are so many buildings slowly being knocked down.

This approach will ensure you remember that this is ancillary information and not the words of your participants. There are, of course, other methods you might implement, but whichever you choose, ensure it is adopted consistently throughout the transcripts and that this information should comprise a level of detail that is specific to the data being produced and the scope of the research questions you have established.

Representing 'the voice' on the page

A final point on transcribing relates to how precisely you transcribe the spoken information on the page. Gibbs (2007: 11) argues that there are "dangers when moving from the spoken context of an interview to the typed transcript" in that, given the change in medium from spoken to written words, elements of the interview may become lost. In contrast, others, such as Clark et al. (2019: 218) contest, that "[t]he spoken word does not always lend itself to the written word", meaning we do not often speak in the same manner or pattern as we write. Hence, there are opposing schools of thought on what constitutes verbatim transcription and how far a transcriber should go in representing the words of a participant on the page. Some argue for including every 'um', 'ah' and 'like' that exists in the recording, as well as including considerable detail on indications of pauses, interruptions, hesitations or repetitions. Others weigh up the value of focusing on representing the actual responses to the questions answered by the participant so as to avoid the transcript becoming overly 'noisy'. Some researchers even transcribe in the dialect of the participant to present an honest and accurate depiction of the spoken word that reflects their social and cultural identities (Macaulay, 1991).

These approaches may seem a little fussy and unnecessary to the novice researcher, yet understanding how your participants have responded to your questions may be just as valuable to your research as the spoken words they provide. It is therefore vitally important not to adopt selective methods of transcription (i.e., miss out on seemingly unimportant information or sweep through your data for quotes). This would be considered poor research practice – more akin to a journalistic approach to handling interviews – and is likely to steer the analysis away from the participants' responses towards forcing the data to 'prove' answers to the research problem.

It is important to remember, though, that transcripts are merely interpretations of the spoken word. Spoken language and written language each have different rules, meaning it is impossible for a transcript to be truthfully representative of the interview encounter. For example, Brinkmann and Kvale (2018) observe that it may be difficult to accurately determine, or represent, the differences between a pause and the end of a sentence. There may well also be instances whereby the partial transcription of an interview recording is necessary – you might, for example, have identified a portion of the interview recording that does not align with the objectives of the research. Transcribing the entire interview is, of course, the most complete approach, but limited time and resources can sometimes mean transcribing is achieved pragmatically (see Halcomb and Davidson (2006) for a critical discussion of the pros and cons of verbatim transcription). I would advise anyone transcribing partially to make sure they refamiliarise themselves with the audio recordings during the analysis process to ensure key information and context are not lost.

WHAT IS INTERVIEW DATA ANALYSIS?

Gibbs (2007: 1) suggests that analysis constitutes the transformation of data from a broad, messy collection of interview transcripts into "a clear, understandable, insightful, trustworthy and even original analysis". In its most basic form, data analysis primarily involves making sense of your data in order to help respond to the research questions set out in your project proposal. In interviews, this may take different forms, and some research will require the content of the interview transcripts to be the central component of the analysis, while others may involve deeper engagement with the ways in which participants present their own narratives or stories in their responses, as well as how spoken content fits alongside the environments the interviews were conducted in (Rubin and Rubin, 2011; Silverman, 2015; Jarvinen and Mik-Meyer, 2020). Common though among all of these approaches is a need to move quickly beyond a simple description of what has been said to deeper critical engagement with how it has been said and why in order to produce new insights and information. As with all research, the choice of analytical approach should be relevant to the type(s) of data that you have, meaning there is no 'recipe book' or 'instruction manual' for qualitative interview data analysis (Roulston, 2014).

What is important is that interview data analysis is not a linear process. Instead, such analysis is an iterative process in that it requires the researcher to move back and forth through the data, re-reading the transcripts and coding to robustly understand the key themes – and the meanings behind those themes – from the data.[2] This can sometimes be a little bewildering, and some elements of the analysis will inevitably not make it into the final written output, but it is this iterative process of reading and re-reading the transcripts that ensures the most relevant and important themes are identified in a rigorous and robust way.

Coding interview data

Coding is the starting point for most interview analysis and involves sorting, categorising and indexing the data within interview transcripts (Gibbs, 2007) in order to condense data and aid interpretation (Brinkmann and Kvale, 2018). This section briefly describes what coding is by outlining the differences between a priori coding approaches (i.e., commencing the analysis using pre-determined codes) and open coding approaches (i.e., allowing themes and ideas to emerge from a close reading of the text). I have presented an ano-nymised hypothetical example to indicate what your coding might look like. Moreover, I will introduce you to axial coding approaches that encourage deeper engagement with data and help refine analysis through iterative, or repeated, reading of the text.

Coding approaches – the basics

In its simplest form, coding is a process of extracting meaning from data in a systematic fashion. As Seale (1999) suggests, coding should be considered the preliminary stage of analysis and involves the researcher reading through and marking up the interview transcripts with codes that denote certain types of words, phrases or themes for later analysis. Coding therefore requires a close reading of *all* of the data, meaning you will need to work systematically through your transcripts, reading them carefully and entirely to identify themes, relationships, contrasts and contradictions between the responses provided by your participants (Robson, 2002). We normally call this first stage of coding 'open coding' in that you read transcripts with an open mind to the data contained within them, and this is where ideas can start to emerge about the relevant topics or themes discussed in your interviews and contained within your research questions.

Coding can be approached in two ways. Traditional coding approaches involve highlighting key fragments of the text in a tran-script and noting pertinent information on the page. This can be achieved either by hand, using highlighter pens, or via a word pro-cessing package – using the highlighter functions (see Figure 9.2). More contemporary approaches involve utilising computer packages that allow the researcher to create code files that information can be dragged and dropped into. For the rest of this section, we will focus on the traditional approach, and in the final section of this chapter, I will introduce you to some of the data analysis software packages.

Crang (2005) refers to coding as the creation of 'theoretical memos' that provide the building blocks for the deeper analysis that will emerge later on. Hence, coding is about identifying themes within a single interview and then exploring the occurrences of these themes laterally across the full set of interviews. As mentioned

Sarah: Yes, but then I suppose it's a social thing. Their hour in-between, their catching up with people. Whereas I think, if I do this now, I haven't got to do it later so, I'm not being rude, but I don't want to sit and talk (laughs). Sometimes I do. I wouldn't be rude to anyone, but I'd rather make better use of my time. And whilst the Hub is not brilliant for studying, it's not worth me going to library for an hour and back again, so I don't do it. But I do tend to sit and read something. I've always got spare stuff that I could be working on so yeah, I'll work when I've got gaps. And sometimes things will get cancelled when you're in so I tend to make use of that time as well.

Mark: Is this quite a solitary task?

Sarah: Yes. I don't think everyone has got the same work ethic as mature students (laughs).

Mark: You said that you have made connections with some younger undergraduate colleagues. How did you establish that?

Sarah: Because I'm doing [subject], one of them is doing exactly the same as me and we're in a lot of the same things. They do call me mum (laughs) and they do ask for my work (laughs). I do give it to them after, when it's been marked. I do think it's quite good, I do get quite good marks and they just want to learn from me. So that's where it started, they started to see what I got and they thought "oh, I think I'll be friends with you". I don't mind that. It's good to be shallow sometimes. I have no thoughts that these friendships will go on or go outside. I don't socialise with them. They have asked me a couple of times, but they are just being polite. They don't really want to go out with me (laughs) and I don't really want to go out with them, but they're funny and they remind me of being nineteen so.

Mark: You're fitting in without really fitting in?

Sarah: Yeah, I don't really want to fit in, I have no real desire to be nineteen again.

Coding Key: Time   Friendships

Figure 9.2 An example of coding using MS Word

earlier, coding operates in an iterative way, meaning interview transcripts should not be coded just once. Clarity of themes and the importance of those themes will only likely emerge through repeated readings of the transcripts, meaning coding schemes will likely adapt and change with each reading until the core themes have been established. You may, for instance, find that with each read, new codes might be added while others that are less significant are removed. Some codes might be collapsed into a simpler code, while other codes become more complex and warrant splitting up into sub-codes. Being open to adaptation through repeated reading is an essential part of the analysis process and ensures that you, as the researcher, remain firmly in tune with the findings produced within the data so that nothing is missed.

Top-down or bottom-up coding approaches

While open coding is usually the primary coding approach adopted by qualitative researchers, other top-down (a priori coding) and bottom-up (grounded theory) methods can also be useful in commencing the coding process, and it is important to consider if, how and why a particular approach might be valuable for your specific research design (see Box 9.2).

A priori coding is considered a top-down approach, whereby themes and codes are pre-determined from the aims and objectives set out in the research proposal. 'A priori' refers to something that can be known without experience and therefore entails looking to verify or refute themes that are anticipated through the interview process. An a priori approach would be considered appropriate when conducting structured or semi-structured interviews, as these are likely to have been designed around a specific set of research objectives. This would be deemed a more investigative process and would be less likely to engage with new findings. Proponents of a priori techniques argue that research should be conducted through structured evaluation in order to test or justify judgements.

An alternative, more bottom-up approach is *grounded theory*. Pioneered in the 1960s by Glaser and Strauss (1967), grounded theory is a technique where themes and hypotheses are generated from the field data in order to construct theories and concepts (Robson, 2002). Grounded theorists therefore start with the data and then

---

Box 9.2   Approaches to coding interviews

A priori coding

- This is a top-down approach.
- Codes are usually themes that can be known without experience.
- Themes and/or codes are pre-determined and are usually derived from the research aims and objectives.

Grounded theory

- This is a bottom-up approach.
- Themes are drawn from reading the data to construct theories and concepts.
- Useful in novel research when theories are unclear or non-existent.

tease out the theory through an iterative process of careful coding and sorting (Charmaz, 2006). This technique is closer to the open coding approach mentioned earlier in this section and is very useful in novel research when theories are unclear or non-existent. However, it is important to recognise that Glaser and Strauss' (1967) emphasis that no preconceived ideas should be applied to grounded theory may have been misinterpreted and that avoiding "preconceived bias, dogma and mental baggage" (Allan, 2003: 8) which might influence or manipulate the interview *process* might actually have been closer to what they proposed.

'Doing' coding

Beyond theorising 'about' coding, it is important to understand what 'doing' coding entails. The transcription process has built towards this point, and the care and diligence you have put into creating clear and readable texts will help you work through the coding process. I identify five key ingredients for coding your transcripts. These are not exhaustive and will probably require several reads of the text to cover completely, but doing this will not only help you establish the core themes for analysis but also how these might translate to the final written document.

Highlight direct quotes to use in written analysis: As discussed throughout this section, working through the transcripts and highlighting pertinent themes is the primary activity involved in coding. The information highlighted will help you understand how your themes work, what relationships and differences exist between how your participants engage with a particular theme and, fundamentally, why this matters in terms of your contributions to knowledge.

Refer to key concepts: An essential part of any research is to relate your findings to existing theories and concepts in your field. Academic research needs to matter so establishing how your findings contribute to knowledge will add necessary impact to your research. You will have carried out preliminary reading in a literature review so reflect back on your notes or reviews while you read your transcripts and consider what about your research is new or original. Try noting down where these instances appear in your research to make it easier to locate the contributions as you begin to analyse away from the transcripts themselves.

Refer to key texts/connections to reading to follow up: As noted earlier, if you are identifying new themes that advance existing knowledge, try to consider the key texts and authors that have developed these theories. Making notes in the transcription will prompt re-reading or further reading to help confirm any original ideas you might be formulating.

Refer to other interviews with similar key points/connected themes: This is a vital step and will help significantly when working laterally across your data. As you become more engrossed in your data, you will start to remember if particular points arise across other interviews or instances where contradictory responses are given. Noting these in the transcript can aid with formulating 'why' a phenomenon occurs and will ensure your arguments are balanced and critical.

Reflect on methodology/positionality: This might sound like a pointless exercise at the analysis point of the research, as the data has already been collected. Reflecting on how the data has been collected is, however, very useful when writing methodology chapters/sections or when formulating subsequent research designs. You might, through your analysis, start to see how your own practice has influenced the responses. For example, might you need to ask more open-ended questions? Do you interrupt your participants when they speak or lose control of the interview process at times? It is often very difficult to reflect 'in the moment', but these types of issues will undoubtedly arise through reading the transcripts through.

Axial coding

Part of the iterative nature of coding is the splitting and combining of codes. As noted earlier, the more you read your transcripts, the more critical you will become of the information presented with them. Axial coding therefore extends open coding by breaking down a large category (e.g., home) into smaller, more nuanced codes (e.g., house, family, domestic activities, pets, children). This is part of the production of a more refined coding scheme (Strauss and Corbin, 1997; Ezzy, 2002) and involves revising the codes as you analyse. Codes therefore should not order or simplify your narrative or data but rather work to *order your thinking*. As Crang (2005: 224) warns, "[C]odes are not there to be rigidly reproduced, not to be counted, but as an aid to the researcher in making sense of the material. They are not an end in themselves". Hence, being flexible and open to adapting your strategy according to the types of information

Q. What are the advantages and disadvantages of Genetically Modified food?

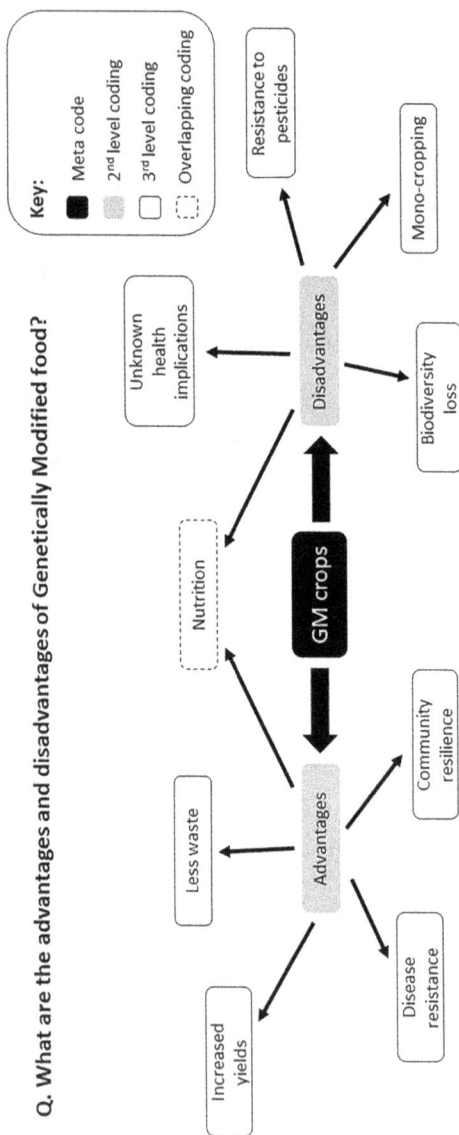

Figure 9.3 An example of axial coding

found within your transcripts is more likely to be a better interpretation of the issues, perceptions and experiences that matter to them.

In Figure 9.3, I illustrate an approach to axial coding that relates to a hypothetical research question relating to the relative advantages and disadvantages of genetically modified (GM) food. In the following, I set these out as a series of steps:

1. When reading my interview transcripts, I have identified 'GM crops' as an important theme to look for (identified as the 'meta code' (or primary code) in the black box). I therefore note every time GM crops were mentioned in the interviews.
2. This will likely generate a lot of information, so to refine this I then break this code into 'advantages' and 'disadvantages' (identified as 'second-level codes' in the grey boxes). This helps sort the coded information according to the characteristics of the question.
3. Yet, these sub-themes might still be rather large and contain contrasting information within them or material that overlapped the sub-themes. An additional set of codes (identified as 'third-level codes' in the white boxes) helps organise the data into more refined sets that identify what the relative advantages and disadvantages were, as well as highlighting whether certain codes were prominent – or in the case of the dotted box, where overlaps occurred.

Axial coding is therefore an important method for establishing more critical analysis of interview data, as it helps you interrogate your coding more thoroughly and robustly.

Transforming codes into analysis

Once the coding process is complete, the next phase of analysis can begin. You will need to find a way of consolidating your coded information into new files or 'piles', and this will comprise splitting up the original transcripts (Gibbs, 2007). As this will involve 'snipping up' the transcripts, always ensure you note which interview the information comes from to ensure you can attribute it to a participant. This is the point at which coded material can quickly lose its original context, and therefore it's meaning, so it is important to have original 'clean' versions of your transcripts available for you to refer back to. The following is an example of a piece of verbatim interview transcript from an interview carried out with a farmer about how he uses music and social media to combat loneliness when out in the fields. The text that includes the code is highlighted in bold.

INTERVIEWER That's quite interesting actually, what you said about being able to listen to your podcasts or your music. Is that a big help when you're out in the field all day?

CHARLIE Yeah, that's a big one. I mean, growing up I listened to the radio. Now, with auto steer [on the tractor] and things like that, it's a bit lazy, but it is quite good that you can sort of keep an eye on other things. And yeah, I think that's a big help. **I think technology, like machinery and music has helped a lot with my loneliness. Also, with technology, if you're scrolling through Facebook all day long, it can get a bit depressing. If you're stuck on your own. Yeah, it probably is good in some ways and bad in other ways, I guess. Yeah.**

When this code was extracted from the transcript and used as a direct quote in a research article, I added an anonymised caption at the end of the quote to ensure I knew which interview it was attributed to:

> I think technology, like machinery and music has helped a lot with my loneliness. Also, with technology, if you're scrolling through Facebook all day long, it can get a bit depressing. If you're stuck on your own. Yeah, it probably is good in some ways and bad in other ways, I guess. Yeah.
>
> (Charlie, 24, mixed dairy and arable farming)

Once you have compiled your new 'code' files, you can then start to read back through these to examine the relationships and contrasts within them. This forms the basis for additional axial coding and will help ensure the analysis is critical and robust. This phase of the analysis process will involve you consistently referring between codes, their content, the original transcripts, the academic literature and any other relevant material to ensure clarity and cohesion. This enables the building and interpretation of ideas.

INTERVIEW DATA ANALYSIS METHODS

It is important to understand that qualitative data analysis is not just about coding, and you may have some very specific ideas about what your interview data needs to say in relation to your research aims and objectives. As with all data analysis, there are fundamental foundations that ensure data are handled carefully, honestly and with integrity (Gubrium and Holstein, 2001; Brinkmann and Kvale, 2018). You may, for example, need to observe the *content* of the data – i.e., the types of words, phrases or inflections that your participants have used during the interview process. Moreover, the *context*, or the ways in which your participants have framed their responses to your questions, may be important in considering why a phenomenon is understood in a certain way. In some instances, you may need to go deeper into the *narratives*, or stories, produced by your participants to analyse depth and meaning. These three approaches are not mutually exclusive and can be used in conjunction with one another to help develop your analysis in greater detail.

Content analysis

The simplest form of qualitative data analysis is content analysis. This is effectively a form of quantification that involves counting the frequency and volume of codes within the transcripts. Content analysis is primarily an objective, systematic and quantitative description of the type and amount of content contained within a set of texts (Brinkmann and Kvale, 2018). It works as a method for studying the meaning, background and intentions contained in a text and achieves this by measuring the attitudes, themes and characteristics of a text by looking for frequencies, patterns, relationships and discrepancies (Robson, 2002). You may be familiar with word clouds as an approach to revealing the relative importance of words relating to a phenomenon. While not a robust form of analysis, word clouds are basically visual representations of content analysis that, if used effectively, can produce a summary picture of the types of words and phrases that have commonly been used in your interviews.

In terms of how you might use content analysis in your own research, say you were trying to understand how residents felt about seasonal flooding in their local community, you might want to start by observing the emotions that your interview participants included in their responses – i.e., looking for the frequencies of words like 'happy', 'sad', 'angry', 'relieved' etc. – to build a picture of what types of feelings are being produced by this issue. To ensure the analysis is robust, you will need to assign rules to a particular code or category. This will enable you to make sure all the interviews are coded consistently (Brinkmann and Kvale, 2018). Content analysis of interview data may, of course, only yield a very broad understanding of a phenomenon, and this approach is often used most appropriately as the foundation, or first step, for subsequent deeper context analysis.

Thematic (context) analysis

Thematic analysis, sometimes also referred to as context analysis, is the most commonly used qualitative analysis technique. This involves shifting the analysis from observing *what* has been said, interpreting and exploring *how* things are said and *why* this matters (Clarke and Braun, 2021). Thematic analysis therefore requires deeper engagement with the data and may well extend beyond the words on the page in order to engage with the non-verbal dimensions of the interview (i.e., the way responses are framed and articulated). A thematic analysis often follows on from content analysis in order to look beyond the frequencies of codes to explore the meaning behind the text. In using this technique, you would be looking for meaning and subtext (the implied meaning of a text that is not announced explicitly in the writing), and there are some key approaches that you might engage with.

First, thematic analysis will require identifying what broad themes, or patterns, are evident across the dataset, but this alone might not tell us *why* these patterns are important. For example, you can present various questions like the following:

- "What were the intentions of the participants in how they replied to my questions?"
- "Were there particular motivations for providing responses?"
- "Did the sequence of questions lead to certain types of knowledge being revealed?"

It is important to consider all of these types of questions when conducting thematic analysis, and this will often mean reading and re-reading your interview transcripts to establish each layer of analysis.

Here is a hypothetical example relating to public perceptions of flood risks. Say a content analysis of interview data has revealed a number of different types of negative emotional language being discussed; we might want to consider why these are significant by examining the context in which they are produced. We could, therefore, adopt a 'bottom-up' approach (commonly referred to as 'inductive analysis') whereby themes are derived from specific knowledge produced during your interviews that then build towards developing a general theory about the phenomenon under investigation (Ezzy, 2002). An example of this might be that there is considerable evidence in the transcripts of participants' feelings towards existing flood mitigation strategies and how these affect their everyday lives. We could then take this theme and deploy axial coding to refine it further and scrutinise how and why these patterns have emerged. For example, we might look for evidence of if, and how, there might be differences in responses relating to things like age, employment type and housing tenure, among other things, to help us elaborate on the points raised by participants.

Narrative analysis

Social scientists are becoming increasingly interested in exploring interview data in new ways that reveal even deeper layers of meaning that thematic analysis alone cannot produce. Detractors of traditional qualitative analysis techniques argue that thematic analysis produces knowledge that is often fragmented and that 'snips away' at the edges of what interview participants have said to focus on what the interviewer deems important rather than what the interview participant has said. This, perhaps, runs the risk of distorting the meaning and intention behind how and why a participant articulates their interview responses (Ezzy, 2002). I emphasised in Chapter 1 that good qualitative interviews are effectively about storytelling. Coding and sorting through interview transcripts can break up, or distort, these stories, and it can be difficult to reassemble them later on during the write-up phase. Narrative analysis is a technique that preserves the stories as whole entities (Bamberg, 2020). As the term 'narrative' implies, the analysis of these data keeps the transcript intact in order for the researcher to explore insights and meanings in how responses are articulated. For example, the ways in which a participant sequences their responses or puts particular emphasis on parts of their story can reveal significant details on how and why a phenomenon exists.

Let us return once more to the flooding example. We have asked our participants to detail their experiences of a particular recent flooding event. In reading through the transcripts, we can start to see patterns emerging in how the participants have prioritised certain events and issues surrounding the flooding phenomenon. Those living close to the flooding site might steer the discussion towards how they were caught out by the inundation of water and the implications for recovering their damaged property. Those further away from the flooding site might focus their narratives on mitigation and moving their property above the inundation levels. Exploring the ways in which these participants describe the event can, for example, tell us something about how flood mitigation strategies need to be nuanced depending on the proximity of a risk site.

DATA ANALYSIS SOFTWARE

This final section will offer a very brief outline of computer-assisted qualitative data analysis software (CAQDAS), alongside its relative strengths and weaknesses, and signposting to appropriate software guides.[3] As I mentioned earlier in the chapter, interview analysis is a long process and can generate significant amounts of written data. This can make the post-interview process time-consuming and unwieldy, particularly for novice researchers. A fairly recent approach to interview data analysis has been the adoption of digital packages that can assist with transcribing, managing and analysing qualitative data. CAQDAS provides opportunities for the researcher to contain interview data in one single place with greater chances for creativity and experimentation in how data is analysed than conventional printed material can often provide (García-Horta and Guerra-Ramos, 2009).

CAQDAS has many benefits for researchers in terms of data management, the flexibility of coding strategies, the speed of data recall, and the audit trail that can be generated through the analysis. It is vital, though, to remember that CAQDAS is a data management tool and cannot do the analysis for you. As with the traditional approaches outlined earlier, you will still need to generate codes, search for the codes in your transcripts, weigh up the value of coded material and break up, or collapse, codes. The CAQDAS will therefore help you manage and visualise the outputs from your own analysis (Carcary, 2011). What CAQDAS is particularly useful for is experimenting with codes. As the package will often allow you to save coded material, the ability to produce new codes, or identify relationships between codes, is highly productive in ways that printing and editing transcripts by hand would never be able to achieve. CAQDAS products are also fairly intuitive, meaning they are easy to learn and can be used simply (e.g., dragging and dropping coded material into the relevant point of a list of codes), or more advanced techniques can be utilised, such as code mapping.

As with all techniques, CAQDAS comes with disadvantages. Software is, of course, no replacement for analytical clarity, and CAQDAS is not capable of running analysis independently (García-Horta and Guerra-Ramos, 2009). As a researcher, you will need to rely on your analysis skills and use the programme to help support the management of your thinking. CAQDAS products also differ in type and quality – with some packages being suited to specific tasks. Likewise, you may be at the mercy of licences, meaning the package available through your institution or organisation might not necessarily be the most appropriate one for you. Different programmes may not match the way you want to work. Some packages will only analyse text, while others might allow you to add images, video or audio to your dataset and cross-reference between them. Moreover, some tools will only allow you to code your transcripts, while others might aggregate your data, permit queries to be run, visualise your coding schemes or even statistically analyse the findings. This all sounds quite exciting, but CAQDAS can be time-consuming to learn and usually requires a minimum number of interviews to be worth using. So, managing your expectations alongside the time you have to analyse the data will help you understand the value of CAQDAS in your project (see Box 9.3). There is now a wide variety of CAQDAS products available, and the University of Surrey has produced an excellent guide to choosing the right package for your project. This includes a guide to choosing the right CAQDAS product and reviews of the leading software available to download: https://www.surrey.ac.uk/computer-assisted-qualitative-data-analysis/resources/choosing-appropriate-caqdas-package

---

Box 9.3 Utilising CAQDAS in analysis design

Silver and Lewins (2014) present a very clear approach to working with CAQDAS in research. They outline four elements of computer-aided analysis that researchers can find easy to work with. These are presented here in a list, but, as with all other approaches to analysis, the researcher may be required to move backwards and forwards through the points, depending on the relative stage of the analysis.

Organise data: CAQDAS can encourage both free coding and more a priori approaches and easily facilitates more granular schemes that group codes into themes, create micro-codes and link codes together. These can often be visualised in maps or diagrams.

Explore data: As well as coding, CAQDAS can allow researchers to annotate their codes and transcripts with further information. Using search tools or query functions can start to reveal patterns within the data that may not be obvious when looking at raw transcripts.

Interpret data: This is when the patterning becomes most important and the point at which the data moves from transcribed material to analysis. The CAQDAS can help establish how themes interconnect by allowing the researcher to move between coded analysis and the original transcripts with ease. Here, connections can be made between codes, and theory can either be tested or generated.

Integrate data: As theory is generated, the CAQDAS can help with the writing and presentation of the data. This might involve integrating qualitative data with other forms of data, like statistics, or helping make connections with the theories and concepts in the academic literature.

SUMMARY POINTS

- Carefully and accurately transcribed interview data will provide a strong foundation for robust analysis and presentation.
- Transcription can also form part of effective and clear data management in terms of data storage and the consistent structuring of transcripts.
- Anonymity and confidentiality must be adhered to when transcribing interviews, meaning names and other identifying information should be removed before analysing data.

- The presentation of participants' voices should be considered carefully to ensure the subsequent analysis is an accurate, or fair, representation of their experiences and perceptions.
- Data analysis constitutes the transformation of a collection of broad interview responses into clear, understandable analysis. It involves making sense of data in order to help respond to the research questions set out in the project proposal.
- Qualitative data coding is a process of extracting meaning from interview data in a systematic fashion to identify specific themes, relationships, contrasts and contradictions in and between responses.
- Coding cannot be completed just once. Robust analysis will require repeated readings of interview transcripts to understand how themes work, what relationships and differences exist between how participants engage with a particular theme, and fundamentally, why this matters.
- Data analysis will usually comprise observing the content of data, examining the context in which responses were framed, and understanding the narratives, or stories, that have been produced through the interview discussions.
- CAQDAS can be supportive in managing data and organising the coding and analysis process but cannot actually analyse the data for you.

WHAT TO DO NEXT?

Consider the following:

- Transcribe your interviews as soon as you can. Avoid waiting until you have completed the data collection, as this will inevitably lead to data fatigue.
- Do not underestimate the time it can take to transcribe. Build plenty of time into your research schedule to do this, even if using transcription software.
- Establish your data analysis strategy very early on in the interview design phase. This will mean that you have a clearer idea of what types of data you are producing and what these are likely to look like once you begin analysing. Avoid any unnecessary surprises.
- In doing the aforementioned, decide on your coding approach – top-down or bottom-up – and what sorts of themes you might be anticipating or looking for in your research.

- Think about how you intend to manage the analysis process. If you are generating lots of data, then perhaps a data analysis software package might help you manage the process efficiently.

Suggested further reading

Brinkmann, S. and Kvale, S. (2018). *Doing interviews*. London: SAGE.
    This book contains a chapter dedicated to handling the post-interview stage of the interview process, including transcribing and analysing interviews.
Gibbs, G. R. (2007). *Analysing qualitative data*. London: SAGE.
    Gibbs provides some very accessible approaches to managing the analysis phase of a research project, including guidance on generating coding schemes.
Hepburn, A., and Bolden, G. (2017). *Transcribing for social research*. London: SAGE.
    This book provides a comprehensive guide for approaching interview transcription, including handling speech and non-verbal cues, as well as organising transcription on the page.
Silverman, D. (2015). *Interpreting qualitative data* (5th ed). London: SAGE.
    In Silverman's book, readers are introduced to a variety of data analysis methods and approaches that guide the reader through the relative strategies to ensure credibility through data analysis and presentation.

NOTES

1 A word of caution. If you take time to carefully anonymise your data, please make sure you do not identify your participants by thanking them directly by name or organisation in any acknowledgements you might include in your final write-up.

2 Indeed, when analysing there is also significant value in valuing what is *not* said in an interview, or what types of questions might *not* have elicited responses. This can help provide context for the analysis in terms of recognising the importance (or not) of certain questions, as well as the motivations of interview participants in responding to questions.

3 Two of the most popular packages are NVivo (https://lumivero.com/products/nvivo/#:~:text=NVivo%20is%20Lumivero's%20easy%2Dto,from%20their%20qualitative%20data%20faster) and ATLAS-ti (https://atlasti.com/). Check with your institution or organisation to find out if you have a licence to use either of these products.

REFERENCES

Allan, G. (2003). A critique of using grounded theory as a research method. *Electronic Journal of Business Research Methods*, *2*(1), 1–10.

Bailey, J. (2008). First steps in qualitative data analysis: Transcribing. *Family Practice*, *25*(2), 127–131.

Bamberg, M. (2020). Narrative analysis: An integrative approach. In Jarvinen, M. and Mik-Meyer, N. (Eds) *Qualitative analysis: Eight approaches for the social sciences* (pp. 243–264). London: SAGE.

Brinkmann, S. and Kvale, S. (2018). *Doing interviews*. London: SAGE.

Carcary, M. (2011). Evidence analysis using CAQDAS: Insights from a qualitative researcher. *Electronic Journal of Business Research Methods*, *9*(1), 10–24.

Charmaz, K. (2006). *Constructing grounded theory: A practical guide through qualitative analysis*. London: SAGE.

Clark, T., Foster, L. and Bryman, A. (2019). *How to do your social research project or dissertation*. London: Oxford University Press.

Clarke, V., and Braun, V. (2021). *Thematic analysis: A practical guide*. London: SAGE.

Cope, M. and Kurtz, H. (2016). Organising, coding and analysing qualitative data. In Clifford, N., Cope, M., Gillespie, T., and French, S. (Eds) *Key methods in geography* (2nd ed., pp. 647–664). London: SAGE.

Crang, M. (2005). Qualitative methods: There is nothing outside the text?. *Progress in human geography*, *29*(2), 225–233.

Ezzy, D. (2002). *Qualitative analysis: Practice and innovation*. London: Routledge.

García-Horta, J. B., and Guerra-Ramos, M. T. (2009). The use of CAQDAS in educational research: Some advantages, limitations and potential risks. *International Journal of Research & Method in Education*, *32*(2), 151–165.

Gibbs, G. R. (2007). *Analysing qualitative data*. London: SAGE.

Glaser, B. G., and Strauss, A. L. (1967). *The discovery of grounded theory: Strategies for qualitative research*. Chicago: Aldine Publishing.

Gubrium, J. F., and Holstein, J. A. (2001). Analytic strategies. In Gubrium, J. F., and Holstein, J. A. (Eds) *Handbook of interview research* (pp. 671–674). London: SAGE.

Halcomb, E. J., and Davidson, P. M. (2006). Is verbatim transcription of interview data always necessary? *Applied Nursing Research*, *19*(1), 38–42.

Hall, S. M., Sou, G. and Pottinger, L. (2021). Ethical considerations in creative research: Design, delivery and dissemination. In von Benzon, N., Holton, M., Wilkinson, C. and Wilkinson, S. (Eds) *Creative methods for human geographers* (pp. 49–60). London: SAGE.

Hepburn, A. and Bolden, G. (2017). *Transcribing for social research*. London: SAGE.

Jarvinen, M. and Mik-Meyer, N. (2020). Analysing qualitative data in social science. In Jarvinen, M. and Mik-Meyer, N. (Eds) *Qualitative analysis: Eight approaches for the social sciences* (pp. 1–27). London: SAGE.

Kvale, S. (2007). *Doing interviews*. London: SAGE.

Macaulay, R. K. (1991). "Coz it izny spelt when they say it": Displaying dialect in writing. *American Speech*, *66*(3), 280–291.

McLellan, E., MacQueen, K. M. and Neidig, J. L. (2003). Beyond the qualitative interview: Data preparation and transcription. *Field methods*, *15*(1), 63–84.

Oliver, D. G., Serovich, J. M. and Mason, T. L. (2005). Constraints and opportunities with interview transcription: Towards reflection in qualitative research. *Social forces*, *84*(2), 1273–1289.

Robson, C. (2002). *Real world research: A resource for social scientists and practitioner-researchers*. Chichester: Wiley-Blackwell.

Roulston, K. (2014). Analysing interviews. In Flick, U. (Ed) *The SAGE handbook of qualitative data analysis* (pp. 297–312). London: SAGE.

Rubin, H. J. and Rubin, I. S. (2011). *Qualitative interviewing: The art of hearing data* (3rd ed). London: SAGE.

Seale, C. (1999). *The quality of qualitative research*. London: SAGE.

Silver, C. and Lewins, A. (2014). *Using software in qualitative research: A step-by-step guide*. London: SAGE.

Silverman, D. (2015). *Interpreting qualitative data* (5th ed). London: SAGE.

Strauss, A. and Corbin, J. M. (1997). *Grounded theory in practice*. London: Sage.

Stuckey, H. L. (2014). The first step in data analysis: Transcribing and managing qualitative research data. *Journal of Social Health and Diabetes*, *2*(01), 6–8.

10

PRESENTING DATA
Putting your participants' words on the page

Chapter objectives

This chapter will explore effective methods of presenting interview data and analysis in written outputs. By reading the chapter, you should

- understand how to transform analysis into written prose;
- appreciate how to represent participants' words effectively and sympathetically in writing;
- develop knowledge of how to summarise, paraphrase and quote from analysis using a range of techniques;
- identify strategies for effective structuring and editing of written analysis; and
- acknowledge if, and when, it might be appropriate to quantify interview data.

INTRODUCTION

In this chapter, I explore effective methods of presenting interview data. This is an important step, as novice researchers can sometimes be confused about what to present (and, conversely, what to leave out) in a final document or feel compelled to transform and illustrate interview data numerically in the same way as survey responses. The write-up phase of your project is ultimately the point at which you get to showcase your findings and start making your contributions to knowledge clear to the reader (Grbich, 2013). Writing up the analysis typically forms one of the largest parts of your dissertation, report or thesis, so it is important that you get this right so as to do

DOI: 10.4324/9781003292784-13

justice to your interview participants' narratives as well as emphasise all of your own hard work.

Yet, transforming potentially thousands of words of analysed interview transcripts into written prose can be daunting, and you may be forgiven for perhaps feeling a little overwhelmed when you reach this stage of the research (see Teow and Holton's (2021) research on documenting grief within a Singaporean family for a discussion of how this can feel in practice). In contrast to quantitative analysis, which can typically be illustrated using graphs, tables or charts, the presentation of qualitative analysis involves synthesising and summarising the findings from across the interview transcripts and then using quoted material to exemplify the key findings of the research (Dunn, 2005). It may not necessarily seem obvious how to do this, and striking a balance between what should be paraphrased and what needs to be quoted can be a tricky process that you may have to practice a few times before getting it right. In this chapter, I will guide you through some vital steps for sorting, organising and presenting your findings in order to help overcome some of the pitfalls associated with clear and effective presentation.

The rest of the chapter is structured as follows. In the first section, I set out some objectives for transforming analysis into written prose, using some useful step-by-step guidance. Next, I provide some critical reflection on how to represent your participants' voices within your written analysis. Third, I guide you through three intersecting approaches to presenting interview data analysis – summarising, paraphrasing and quoting – and how to situate these in your written work. This continues with a discussion of how to consider structuring your writing and the important role of editing in making your writing as clear and polished as it can be. Finally, I briefly touch on quantification in qualitative research and where numbers might, sparingly, be useful.

WHAT AM I TRYING TO ACHIEVE IN PRESENTING INTERVIEW DATA?

In Chapter 9, I set out ways of analysing interview data using specific techniques that aim to understand how the themes that cut across your dataset need to respond to the research aims and objectives of your project. In some respects, this can often feel like the final hurdle of the research process when in actual fact, the presentation of these findings is really the most important aspect (Cloke et al., 2004; White et al., 2003). This might take the form of a dissertation or thesis, a report or paper, or it may be something you present visually, such as a conference paper, a talk or a video presentation. Whichever your chosen (or required) medium, there are some steps involved that are common to the presentation of all interview data. For the purposes of this chapter, I am focussing on 'writing up', but you could use the same principles for other forms of presentation too.

There is often a temptation among new researchers to feel compelled to include every last bit of analysis in the writing, and in doing so, the discussion can become thin and descriptive. The art of writing up analysis is to present the arguments that most effectively respond to the overarching aims and objectives of the research in the clearest and most persuasive way (Rubin and Rubin, 2011). This can be difficult and often requires a few attempts at drafting before things start to fall into place. Try not to panic, though, as this is normal, and even the most seasoned researcher will have to think carefully about how best to present their arguments and may well attempt various iterations of analysis write up before getting it right.

To help you get ready to start your write-up, I set out a few objectives for you to follow. This list is not exhaustive but following this advice will help you get to a point in which you can start transferring your analysis into longer-form prose.

Identify key themes in your coding that relate to your objectives

Now, you may already have completed this task during the analysis phase, but it is important to ensure that the information you have found out from your analysis aligns with the objectives you initially set out to investigate. The coding schema that emerged from your analysis may be fairly broad, and you might find that several themes have emerged from the data that relate to the overall project aim in different ways. At this stage, it is important to consider the relative value of each theme to the objectives set. You may, for example, have found out something very interesting from the dataset but it sits outside the remit of an objective or research question. Moreover, you may have an interesting theme, but it perhaps only draws on a very small part of the data and is therefore difficult to infer meaning from. In these instances, focussing on the themes that respond directly to the objectives of the project can help narrow the analysis down to its core findings and therefore support lengthier, more detailed discussion.

Explore themes across the dataset (looking for relationships/contrasts)

This sounds obvious, but it is important to check that your analysis is a fair reflection of the sample of participants that you interviewed. Of course, there may be instances where themes perhaps come from a sample from within the data (e.g., due to demographic or spatial differences) but be conscious of how representative your coding is of the sample as a whole. Look to see whether the themes you intend to draw upon in your write-up are balanced, both in terms of content (i.e., the volume of coded material within each theme) and coverage (i.e., the range of material across the dataset). This has two benefits. First, understanding the volume of material you intend to work with will enable you to write sections or chapters of comparable length. Providing one very long section and two very short ones can weaken the significance of the arguments being presented. Second, excellent analysis derives from balanced and critical arguments, so having analysis that presents a range of viewpoints can help demonstrate deeper, more sophisticated thinking than simply presenting one argument.

Reflect responses from a range of participants/sources

Again, ensuring that the themes within the analysis are representative of a range of participants' experiences is also very important, particularly when choosing quoted material to include in your writing. It can sometimes be tempting to draw from the participants with whom you perhaps bonded most or who just spoke really well about the topic and put less emphasis on others that perhaps shared less information or did not deliver that 'killer line'. Thody (2006) argues for 'polyvocality' as an approach for showcasing the multiple and varied voices and sources that comprise your interview dataset. Hence, taking a holistic approach to ensure your participants are represented fairly across the writing is a crucial step in maintaining rigour and responsibility in qualitative research. This might not necessarily mean you have to pick quotes from every participant, but it will mean you need to ensure your participants' voices and experiences are represented in the more general analysis. A useful approach to take here is to produce a simple map that illustrates whom the main summary points and quotes are attributed to. This could take the form of a mind map, flow chart or even just a list of bullet points; however you do it, visualising the data can help reveal how you are using it and whose voices you are representing in the final write-up.

Consider demographics/participants' backgrounds as potential drivers for responses

Remember that when analysing interview data, the characteristics of the sample will play a significant role in how participants might respond to questions. By the time you get to writing everything up, it can be easy to get to a point where you might forget that the coded material was articulated by actual people. Before you start to write, take a bit of time to go back over your original transcripts to re-familiarise yourself with your participants again. Think about how their age, gender, ethnicity, etc., might have led to their responses being coded in certain ways. Question whether this makes a difference to how a code has been derived (e.g., perhaps some younger participants gave a different response to a question than, say, older participants. Why might that be?), and consider if and how this needs to be accounted for in the write-up.

Consider emotional/non-verbal cues

As with demographic information, the non-verbal cues that exist in your interviews may well tell you a lot about how and why a participant responded to a question in the way they did. Like the aforementioned example, this information can quickly become lost when interviews are coded, so scan back through your transcripts and audio recordings to see if any additional information needs to be included in your writing. For example, you may find it beneficial to tell the reader how your participants responded to a particular question. Did some participants pause and ponder their responses, while others jumped straight in with an answer? Perhaps, if you asked participants to recount a certain type of event, they might have framed their stories in particular ways, in terms of the tempo of their speech, the language they used, or the relative detail of the examples included (see Box 10.1)?

> **Box 10.1   Things to avoid when writing up**
>
> As well as forming a checklist of things to do, there are also certain things that you will need to avoid doing when moving into the write-up phase.
>
> - There may be a temptation to consider quantifying your data – for example, reducing the themes within your data into percentages. I will talk about this more later in the chapter, and there can be value in illustrating the size and shape of your data, but it is important to remember that qualitative approaches require a depth of explanation of data rather than being concerned with making broad, generalisable points.
> - Another issue to avoid is to skip over the analysis stage and simply look for quotes in the data to present in your writing. It is important to analyse all of the data properly so as to understand fully how the interview participants' responses relate to one another and ultimately relate to the project's aim and objectives. Mining the transcripts for quotes would therefore be considered a more journalistic approach rather than a scientific one; hence, the quotes you use should illustrate or exemplify the themes you have established from your thorough analysis of the data.
> - Sometimes there can be a temptation to 'clean up' your quotes to make them more readable or to emphasise the point you want the quote to make. This means manipulating your participants' words and would not be considered scholarly (see Box 10.3 for advice on how to edit a lengthy quote correctly).

TELLING 'THEIR' STORY HOW IT IS?

In this section, I will outline the importance of retaining participants' voices in your writing. This links back to the earlier discussion about how to represent the participants through the transcribing process and ultimately should remind you of the value of returning to the interview transcripts and audio recordings to ensure the stories being told are those that the participants spoke about.

As with the discussion of transcript content in Chapter 9, there is much contestation in the academic literature as to the extent to which interview extracts should be presented verbatim in the final write-up of analysis – that is, if the 'um's, the 'ah's and pauses are included, as well as whether the accent and dialect of a participant should be accounted for in the text. Brinkmann and Kvale (2018) argue that interview extracts ought to be coherently written in that while their extraction from the transcript might make sense to the researcher, presenting them in a raw form may render them incomprehensible to the reader. They point to writing in the vernacular (the language or dialect of the speaker) as particularly problematic. Brinkmann and Kvale (ibid.) go on to argue that presenting verbatim conversational text or accenting speech could devalue the importance of the analysis and even risk offending or stigmatising participants if they are not expecting to view their speech in that context. That said, others believe that verbatim presentation helps to present a more rounded context to a participant's response that editing might risk losing, and Macaulay (1991) suggests taking a common-sense approach to this.

USING INTERVIEW DATA – SUMMARISING, PARAPHRASING AND QUOTING

One of the trickier tasks to get right can be balancing the broader discussion that emerges from your analysis alongside the quotes that you include in the text that exemplify this discussion. The art of presenting analysis is to summarise the general points, themes and perspectives that emerge across a set of interviews and then embed quoted material that contextualises and authenticates the narratives presented. As with all writing, balance is key to achieving a clear and well-rounded analysis chapter. For example, lengthy summarising with little or no quoted material can make it difficult for the reader to pinpoint how the analysis has been derived from the data. Conversely, too many quotes can disrupt the flow of the writing and make it hard to understand how the material links together. There is no definitive method here, but I outline three approaches for achieving this in the following sections, based on *summarising*, *paraphrasing* and *quoting* that can help you present your findings in a clear, critical and meaningful way.

Summarising

The first step to a successful write-up will be to summarise the broader themes from your findings. Summarising means taking the most important and valuable points from your data and conveying them in your own words. This might involve summarising a theme from the data analysis in just a few sentences to help set out the broader findings from the analysis.

For example, you might want to open a section or chapter with a brief synopsis of the theme you are presenting and how this relates to the research objective or gap you are addressing. This will involve describing the context of the theme, its relative characteristics and how these relate to the extant concepts and theories found in your reading. Summarising can be a valuable exercise in starting the writing process, as this often helps you consider your analysis in the broadest possible way before you then delve into the deeper critical discussion as you write.

Paraphrasing

Alongside summarising, you will also need to paraphrase your participants' responses in your write-up. While summarising allows you to consider the breadth of your themes, paraphrasing encourages you to focus your argument by considering the detailed characteristics of the theme, how these relate to one another, and what contrasts might exist – basically, what does the theme *actually* mean? Paraphrasing works in a similar way to summarising in that you are articulating the analysis of your interviews into your own words to describe and explain the phenomenon under investigation (see Box 10.2). Where paraphrasing differs from summarising, though, is that paraphrasing includes more detail and context.

Box 10.2   Approaches to paraphrasing

Let us use an example of a hypothetical project to illustrate how to paraphrase. You are investigating young people's political affiliations, and one of your objectives involves examining whether these associations change as young people mature and potentially leave home.

One finding you might report is that these associations could be influenced by young people leaving home for work or education. This would involve looking for *relationships within the entire dataset* and explaining how these illustrate the broad patterns of mobility (e.g., it might be common among your participants to infer that they emphasised certain political opinions when they were around new friends or perhaps supressed them when they visited family).

Yet, you might also have exposed differences within the dataset between the young participants' living circumstances, meaning you may need to distil down to *focus on specific subsets of the data* to illustrate contrasting experiences and why these matter (e.g., while all participants might mention moderating their political opinions when they return home, there might be differences between how such affiliations are articulated between young people that have left home permanently and those living away temporarily).

Finally, there might also be specific accounts that exemplify this phenomenon that you could draw upon. This requires you to *outline the specific experiences* of one or two participants and will likely prompt you to present some quoted material to support this (e.g., a specific example might relate to a participant who spoke in their account of the difficulties they faced in expressing their newly formed political opinions with their family members when they temporarily returned home during a vacation period).

In order to respond to an objective, you may need to provide context for your participants' experiences to illustrate how they relate to, or engage with, a phenomenon. This might involve describing a salient characteristic that emerged across all, or part, of your dataset that helps respond to an objective. Whichever approach you choose to take here will involve describing the details of a theme in depth and explaining how this relates to the theories and concepts found in the academic literature.

Quoting

Using direct quotes involves presenting the exact words from a transcript to exemplify the discussion that emerges from your paraphrasing. Quotes are essentially a form of evidence that supports your analysis, and their inclusion is a vital stage in your write-up. There is no blueprint for how to quote, what to include or indeed whether we should quote interviews at all (see Bissell, 2023). I urge you, therefore, to have a look at how others do this in the books and articles you read (see Box 10.3).

For example, some researchers might present a quote first and then use paraphrasing to explain how the themes found within this quote relate to the research aim and objectives. This can be useful if you need to provide context for how or why a participant has made a certain point (i.e., "as XXXX states in the above quote…"). As this is a more linear process, it can also make it easier to structure critical discussion into manageable chunks.

Box 10.3   Presenting quotes in your write-up

Below are some hints to help you present quoted material in your writing. There is no universal system to this, so if you have briefing instructions, please check for any specific requirements before you start.

- Always use quotation or speech marks around quotes to identify the quote as separate from your summarised or paraphrased text.
- Do not change or modify the text within a quote without acknowledging the edit. If the quote needs to be shortened, then denote where this edit has taken place in the text using ellipsis points contained within square brackets, like this: […]. This symbol can also be used to illustrate pauses or breaks in the interview. Examples of editing might be if there is a lot of extraneous information that potentially gets in the way of the main thrust of the quote or if the quote contains information that might identify a participant.
- You might feel compelled to try and emphasise your quotes using italicised or bolded text. This can be quite distracting for the reader and can sometimes add emphasis where it is not needed. If you want to draw attention to a specific point within the text where the participant has emphasised something (e.g., the tone of speech briefly changed or a point was exclaimed), then you can put the word, or words, into italics and then include 'emphasis added' in the caption at the end of the quote.
- Your quotes are important and illustrate the findings from your data, so it is important that you accentuate where they exist in the writing. Try putting your quotes on a new line and indent the whole quote from the left margin (select the text and press 'tab'

on the keyboard). This will make the quote stand out more clearly from the rest of the summarised and paraphrased text.

- Another important element to include is a caption for each of your quotes. These should usually be situated at the end of your quote, between the quotation mark and the full stop, and need to contain some information that helps illustrate where the quote has come from. If you have used a pseudonym for your participant then this will usefully link it to the participant the quote is attributed to. Moreover, there might be important information relevant to the project that you could include, such as age, gender, location or employment status. However you choose to assemble your quotes, make sure that you are consistent.
- Try to use multiple quotes from a range of participants to help demonstrate the breadth of the theme you are writing about. This helps the reader understand that the theme is important enough to be mentioned by a range of sources, as well as illustrating nuance in how the theme is discussed by your participants.

Other researchers might provide the paraphrasing first, by way of describing the participants' experiences and then place the quote after this as evidence. This approach is useful if you want to exemplify the discussion you have made (i.e., "these points are articulated here by XXXX…") and can allow more flexibility in providing more than one quote to illustrate relationships within the data. Often, this approach will require a little further explanation to be included after the quote to tie the paraphrasing and quotes together. I have set out some key pointers in Box 10.3 that can help you achieve the right results.

STRUCTURING YOUR DISCUSSION – ITERATIVE APPROACHES TO WRITING

So far, I have guided you through the stylistic approaches of academic writing. Understanding *what* to write about is important but knowing *how* to write is equally as important. Hence, much of your time writing will be spent tailoring the structure of your discussion chapter or chapters. Unlike standard essays or reports that often have more linear styles of writing, data analysis – particularly interview data analysis – will usually require you to generate your own plan or structure that is specific to your data and research objectives. Good discussion chapters are therefore rarely written in one single draft. The best work is often the product of lots of careful thought about how best to structure and present the analysis, sequence of points and themes, attentiveness to research objectives and literature gaps and clarity and persuasiveness of the arguments being raised. In this section, I advocate two approaches. First, to prioritise structure early on in the writing process, and second, to acknowledge that writing is an iterative process of drafting, editing and refining your work over time.

Structuring your writing

In terms of structure, you may have a project brief that provides you with a specific approach to framing your analysis. If this is the case, then make sure you follow those instructions carefully. More often than not, though, such guidance is indicative, meaning there is flexibility in how you can structure the work, so it is important to think about how you need to articulate your analysis early on in the writing up process. There are a number of ways of approaching this – for example, using visual techniques like mind mapping, but the key message here is to start getting information on the page as early as possible so as to commit yourself to the writing process. I will caveat this by saying that I am a very inefficient writer, and I actually have several versions of this chapter alone in a file on my computer, but getting going with the writing as early as possible really is the only way you can judge the relative clarity of how you articulate your analysis.

Let us work through an example of an approach. Try opening a new word processing document or notepad page and set out a list of section headings from which to work. Some of these will be obvious – you will need a chapter title, as well as introduction and summary sections – but the main body of the analysis could be sub-divided in many ways, perhaps according to your research objectives or by the key themes drawn from your analysis. These approaches are popular in many academic studies and technical reports and mean that the analysis aligns closely with the objectives of the project. Moreover, a clear structure can help keep you on track, preventing you from including superfluous analysis that strays from the remit of the project.

From here on in, you can then start populating these sections with information from your analysis. Start by using bullet points to list the key findings and then consider their order. Think whether your points might be discussed sequentially (i.e., one naturally leads on to another) or whether you are presenting a set of contrasting viewpoints (i.e., building an argument and then contrasting this with a counterargument). Once you have a loose structure, you can then start building on these points using the summarising, paraphrasing and quoting framework outlined earlier in this chapter.

Editing your writing

The other point I have alluded to in this section is concerned with editing. I will stress again that all good analysis and discussion chapters derive from carefully edited writing, and it is the iterative process of editing – of returning to your writing and revising the text – that hones the arguments being proposed. Editing broadly means raising the standard of a piece of writing by revisiting and revising the content and structure (Charmaz, 2020). This will involve examining the clarity of how the writing is expressed, as well as the coherence of the arguments you raise. Hence, editing is more than just proofreading and polishing work. Proofreading is, of course, a very important part of the writing process, but this should be viewed as separate from editing and one of the last jobs to complete.

To be a good editor is to look simultaneously at the specific detail of the writing but also to be able to step back and examine how the writing as a complete text hangs together and how this affects the flow of the narrative that is being produced. So, to edit successfully, you will need to read your writing with a critical mind. This may require you to be objective – basically, to try and read as neutrally as possible so as to avoid dwelling on how long it took you to write a particular section and instead consider the value and quality of the writing in achieving the goal of responding to your research aim and objectives. In literary circles, this has the ominous title of 'killing your darlings', whereby characters, storylines or scenarios are cut from the writing in order to strengthen the overall thrust of the story arc. In academic writing, editing is much the same, and there may well be a piece of analysis that you have spent considerable time working on that, upon further reading in the context of the whole chapter, you can see gets in the way of the overall argument you are trying to make or simply just does not fit with the rest of the analysis and discussion. Having the confidence to edit this out will improve your writing and strengthen the clarity of the arguments you are making.

That said, I would caution against jettisoning edited text altogether though. In the past, I have given my writing a brutal edit only to discover further down the line that I could have used some of that deleted text elsewhere. Saving multiple drafts of your writing can give you a bit more flexibility to make changes without completely committing to the delete button. If you are creating multiple drafts of work, make sure you label these correctly (e.g., Draft_1, v1) so that you know where to find information but, more importantly, which is the most recent, or final, draft to be worked on or submitted.

Lastly, editing does not have to be your sole responsibility. Drawing on the kindness of friends, family and colleagues can help provide different, often non-expert, viewpoints on your work, particularly as these individuals will be detached from the writing in ways that you cannot be. I suggest providing potential editors and proofreaders with specific tasks. Some may be good at looking at the robustness of your arguments, others could helpfully check the flow and clarity of the writing, while some have a forensic eye for specific details. Remember, not all recommendations will be valuable, and it is your job to sift through these and decide on whether to action them or not, but having a range of opinions may well help you refine elements of the writing that you have been struggling with developing.

MORE THAN WORDS – WHEN TO QUANTIFY

Throughout this book, I have championed the subjective and contextual virtues of working within qualitative interview research, particularly in relation to the stories and narratives that interviews can produce. Yet there may be points within the analysis and write-up where you feel compelled to produce some numbers. While I would discourage you from completely quantifying your interview data, there may be circumstances when illustrating the volume of responses for a certain theme to demonstrate its relative importance in the analysis might be necessary (Chi, 1997) or when the saturation point is perceived to have been met (Lowe et al., 2018). Quantification can be used in three ways.

First, setting out numbers can usefully define and justify the demographic composition of your sample. You could, for example, include a table or paragraph prior to your analysis that illustrates the breakdown of your sample according to demographic characteristics (e.g., age, gender, location, occupation etc.). This can help identify any imbalances within the sample and account for whether or not these influence the analysis. In the following, I include two examples of these tables. Table 10.1 shows how participants' characteristics can be grouped together into indicative categories. Depending on the size of the sample, these can be expressed as totals (n) or percentages, but be aware that percentages of very small numbers can be misleading without any indication of the overall sample size they derive from. This approach can be useful in preserving the anonymity of the sample, as it is impossible to pick out individual participants from these categories.

Alternatively, Table 10.2 lists participants by their pseudonyms and provides specific identifying information for each one. This may be valuable if there are certain characteristics of the participants

Table 10.1 A summary table of participants' characteristics

Category	Participants (n = 35)
Gender	
Female	14
Male	15
Other	4
Not stated	2
Age	
−21	4
22–29	22
30 +	9
Highest qualification	
None	0
GCSE	1
A level or equivalent	5
Degree	20
Master's	8
PhD	1
Location	
Plymouth	28
Devon	2
South West	4
United Kingdom	1

Table 10.2 A demographic profile table of a participant group using pseudonyms

Pseudonym	Year	Gender	Age	Term-time residential circumstances		
				Year 1	Year 2	Year 3
Michael	3	Male	−21	Hall	Rented	Rented
Nicola	1	Female	−21	Rented	—	—
Eathan	1	Male	−21	Rented	—	—
Gary	2	Male	−21	Hall	Rented	—
Lucy	2	Female	−21	Hall	Rented	—
Joanna	3	Female	−21	Hall	Rented	With parents

that, when linked together, provide evidence of relationships within the findings. The data displayed in Table 10.2 supposes some links between university students' residential circumstances, their ages and their year of study in how a sense of belonging might emerge and evolve during a degree (Holton and Riley, 2016). While the qualitative analysis draws out the richness of the interview response, the reader can visually trace the broader characteristics of these relationships through the table, and the author can refer back to them in the analysis. Caution must be taken, though, in using this method in terms of preserving the anonymity of participants. This can be achieved by broadening the categories used. Rather than stating a participant's actual age, grouping ages together into decades will illustrate the relative age range of the sample without identifying the participant.

Second, numbers can be usefully written into the analysis itself to help provide broad context to the phenomenon under investigation or to illustrate the relative significance of a theme or opinion. A common starting point for analysis might be to identify the frequencies of key words or phrases using a content analysis approach (see Chapter 9) and presenting this visually using a table, graph or figure. Content analysis can therefore be a useful visual way of numerically identifying key themes within the dataset. A word of caution on this approach is not to assume the frequencies of a particular word to be the truth. Always return to your transcripts to ensure the context in which a frequent word is mentioned is correct – i.e., that the words highlighted are consistent with the intention of the interview participant.

Third, presenting numbers can also be useful if drawing comparisons within an interview dataset. For example, you may be discussing a theme that derives from a sub-set of your sample and want to report the relative percentage of participants in the text. While this might provide some broader context of the relative proportions of your interview sample a particular theme can be attributed to, remember though, that your sample is likely to be very small, so reporting percentages may well conflate their importance. Reporting that 90 per cent of your sample talk positively about a certain activity will have less impact if the sample is only ten people. The key consideration here is not to lose sight of why you are using interview data for your research project. Numbers should be used sparingly and only to provide a general picture of the data. Stating that a certain percentage of participants responded in a particular way to a question might provide surface value in illustrating the importance of the point made; however, for the quality of the interview data to be retained, any descriptive information must also be explained in detail through deeper qualitative discussion.

If you do want to include quantification in your interview data analysis, some qualitative data analysis software packages include functions that produce content analysis that identifies the frequencies of certain words in a single interview or across the entire dataset (see Chapter 9). This might be useful in illustrating the importance of a word or theme (e.g., the frequency of positive or negative emotional words within the dataset), but caution will need to be taken in ensuring that the context of these words is taken into consideration during the analysis.

SUMMARY POINTS

- The art of writing up analysis is to present the arguments that most effectively respond to the overarching aims and objectives of the research in the clearest and most persuasive way.
- Care must be taken in how analysis – particularly direct quotes from interview transcripts – is presented on the page. Verbatim text might convey a response exactly how a participant said it but might be difficult to grasp or even risk offending or stigmatising participants if they are not expecting to view their speech in that context.

- Good analysis presentation summarises the general points, themes and perspectives that emerge across a set of interviews and then embeds quoted material that contextualises and authenticates the narratives presented.
- A clear structure will help with the presentation of the analysis and the sequencing of points and themes. This will allow you to be attentive to the research objectives and literature gaps; and the clarity and persuasiveness of the arguments being raised.
- All writing will need careful editing through revisiting and revising the content and structure. This will involve examining the clarity of how the writing is expressed, as well as the coherence of the arguments raised. Hence, editing is more than proofreading and polishing work.
- While interview data is concerned with the written/spoken word, there are instances where quantification can help support the analysis: either to define and justify the demographic composition of a sample when providing broad context to the phenomenon under investigation or to illustrate the relative significance of a theme or opinion, or if drawing comparisons within an interview dataset.

WHAT TO DO NEXT?

Consider the following:

- Start planning to write-up while in the analysis phase. Setting out a clear structure for your chapters should help you understand how your analysis needs to be presented on the page. This is likely to take a few attempts, so be adaptable.
- Have a look at examples of written analysis to see how your own might look. Academic papers, reports and student dissertations and project papers are excellent starting points.
- Remember to tell your participants' stories. This will involve the careful summarising of data and the presentation of direct quotes that exemplify the points raised. Have a practice at doing this while you analyse.
- Be prepared to edit carefully. You may well over-write your analysis into too many words or not quite manage to get your point across clearly in early drafts. Give yourself plenty of time to write and get feedback from others who are not

necessarily connected to research. This can really help identify the relative strengths and weaknesses of your writing and data communication.

Suggested further reading

Brinkmann, S. (2013). *Qualitative interviewing*. Oxford University Press: Oxford.

Brinkmann outlines effective methods for the successful communication of results, including the reduction, organisation and representation of data findings.

Brinkmann, S., and Kvale, S. (2018). *Doing interviews*. SAGE: London.

The penultimate chapter of this book focusses on techniques for reporting interview data. Of particular interest is advice on how to write for your audience.

Thody, A. (2006). *Writing and presenting research*. SAGE: London.

Thody provides some very clear and concise advice for transforming interview data into written forms.

REFERENCES

Bissell, D. (2023). Questioning quotation: Writing about interview experiences without using quotes. *Area*, *55*(2), 191–196.

Brinkmann, S., and Kvale, S. (2018). *Doing interviews*. London: SAGE.

Charmaz, K. (2020). Grounded theory: Main characteristics. In, Jarvinen, M, and Mik-Meyer, N. (Eds) *Qualitative analysis: Eight approaches for the social sciences* (pp. 195–223). London: SAGE.

Chi, M. T. (1997). Quantifying qualitative analyses of verbal data: A practical guide. *The journal of the learning sciences*, *6*(3), 271–315.

Cloke, P., Cook, I., Crang, P., Goodwin, M., Painter, J., and Philo, C. (2004). *Practising human geography*. London: SAGE.

Dunn, C. E. (2005). Illustrating the report. In R. Flowerdew and D. Martin (Eds) *Methods in Human Geography* (2nd ed, pp. 312–332). Harlow: Pearson.

Grbich, C. (2013). *Qualitative data analysis: An introduction*. London: SAGE

Holton, M., and Riley, M. (2016). Student geographies and homemaking: Personal belonging (s) and identities. *Social & Cultural Geography*, *17*(5), 623–645.

Lowe, A., Norris, A. C., Farris, A. J., and Babbage, D. R. (2018). Quantifying thematic saturation in qualitative data analysis. *Field methods*, *30*(3), 191–207.

Macaulay, R. K. (1991). "Coz it izny spelt when they say it": Displaying dialect in writing. *American Speech*, *66*(3), 280–291.

Rubin, H. J., and Rubin, I. S. (2011). *Qualitative interviewing: The art of hearing data* (3rd ed). London: SAGE.

Teow, P., and Holton, M. (2021). Showcasing creative methods in your dissertation research. In, von Benzon, N., Holton, M., Wilkinson, C., and Wilkinson, S. (Eds) *Creative methods for human geographers*, (pp. 365–378). London: SAGE.

Thody, A. (2006). *Writing and presenting research*. London: SAGE.

White, C., Woodfield, K., and Ritchie, J. (2003). Reporting and presenting qualitative data. In Ritchie, J., and Lewis, J. (Eds) *Qualitative research practice: A guide for social science students and researchers* (pp. 287–293). London: SAGE.

GLOSSARY

Active listening   The act of listening to speech alongside non-verbal cues to aid more empathetic responses.

Analysis   A detailed examination of data in order to understand a complex scenario.

Anonymity   Refers to the obligation to protect an individual's identity from being revealed intentionally or unintentionally.

A priori   The process of looking for pre-determined codes and themes in the data.

Asynchronous   Events, such as interviews, that are not co-ordinated in real time.

Axial coding   Part of the iterative nature of coding generated through the splitting and combining of codes.

Bias   The unfair prejudice for and against a person or group.

CAQDAS   Computer Assisted Qualitative Data Analysis Software. Digital software packages that can assist with transcribing, managing and analysing qualitative data.

Closed question   A question that only elicits a short, simple response, such as 'yes/no'.

Coding   Coding involves sorting, categorising and indexing interview transcripts in order to condense data and aid interpretation.

Concept (theory)   An abstract or generic idea that is generated from the analysis of data.

Confidentiality   Confidentiality concerns the security of information – specifically personal details – provided by participants that must not be shared with others without consent.

Content analysis   A form of quantification that involves counting the frequency and volume of codes within the transcripts.

Context/thematic analysis This involves shifting the analysis from observing what has been said, to interpreting and exploring how things are said and why this matter.

Convenience sampling Sampling whereby the individuals required to participate in the research happen to be most accessible to the researcher.

Cultural Turn A movement between the late 1980s and mid 2000s within the social sciences and humanities whereby culture and meaning became integrated into theory and method, specifically relating to making marginal voices and knowledge heard and understood.

Data The facts and information that are assembled in order to analyse a phenomenon or make decisions.

Data Management Plan A formal document that outlines the strategy for handling data during and after a research project.

Deception The intentional misleading of participants on behalf of the researcher.

Demographics The characteristics of a population or sample, including age, gender, social class, ethnicity etc.

Dialogue A conversation between two or more people.

Discourse A written or spoken form of communication.

Embodied An expression of, or to give tangible or visible form to, an idea, quality or feeling.

Ethics The principles that govern how research should be carried out to protect the dignity, rights and welfare of participants.

Factual questions Questions that elicit simple answers that are either correct or incorrect.

Generalisable The ability for research findings to be applied to larger populations or other situations.

Grounded Theory The process of allowing themes to emerge from the data, rather than presupposing them.

Ice-breaker questions These can involve preliminary, or warm up questions that introduce the interview and collect very general information.

Informed consent The process of providing full information about the research and what a participant's involvement will be, and that participants give consent to taking part.

Interpretive/Interpretivism Questioning the ways in which individuals or cultures create, or interpret, their own version(s) of 'reality'.

Interview guide   A device that establishes and guides the focus of an interview but should not prescribe or overly direct the process.

Interview pilot   An opportunity to test out the interview format, themes and questions before conducting the research.

Iterative   The process of building, refining and improving the research by examining, and learning from, initial research findings.

Marginalised groups   Communities, cultures or groups that experience discrimination and exclusion.

Messy data   Data – primarily qualitative – that is unstructured, uncontrollable and unpredictable. During data collection, this refers to interviews not being conducted in precisely the same way.

Method   The application or procedure of 'doing' research and generating data.

Methodology   The system or principles that are applied to the chosen field of research. This involves detailed explanation of the data collection process – basically, *what* you propose to do in your research and *how* you intend on doing it.

Narrative analysis   A method of interpreting data that have a storied form.

Non-probability sampling   Refers to the approach of sampling that uses non-random ways of selecting groups or individuals to participate in research.

Non-verbal   A form of communication that does not use words, such as body language.

Notetaking   The act of recording information during an interview, usually information separate to the conversation.

Objective research/objectivity   Scientific research that is conducted without influence from the researcher, in term of bias or opinions.

Open coding   Allowing themes and ideas to emerge from close reading of the text or transcript.

Open questions   A type of question that require participants to respond in their own words.

Paraphrase   The process of putting someone else's ideas, experiences and perceptions into your own words.

Photovoice   A method whereby participants take photographs relating to a place, issue or experience and then work with the researcher to examine and explain the images produced in follow up interviews.

Population   Refers to the entire group from which you intend to draw conclusions.

Positionality   The ways in which our identities, social position and culture might shape who we are, what we believe in and how we act.

Positivism/positivist   The philosophical underpinning of science that favours objectivity, truth-seeking, replication and generalisability.

Power   Refers to the subjective ability to wield power, rather than necessarily dominate.

Privacy   Participants' rights to decide what personal information they disclose or withhold in an interview setting.

Probing questions   An open-ended question that encourages participants to reveal more information about a phenomenon.

Pseudonym   The use of a fictitious name or moniker to disguise the identity of a participant or other identifying characteristic.

Purposive sampling   The researcher uses their expertise to select a sample that is most useful to the purposes of the research.

Qualitative   Any data that generates non-numerical information, such as words, images, sounds etc.

Quantitative   Data that are counted and measured in numerical form.

Quota sampling   Defining specific groups within the population to ensure that a balanced sample is met.

Quoting   Involves presenting the exact words from a transcript to exemplify the discussion that emerges from analysis.

Rapport   Developing a close, empathetic and harmonious relationship to help improve communication.

Reflexivity   Acknowledging your role in the research and how your identity might influence the research process.

Replicability   In interview research, this refers to interviewers adhering to responsible approaches when designing, undertaking and analysing interview materials, which ultimately ensure interviews are rigorous and credible.

Research aim and objectives   The aim is a statement of intent, while objectives should be specific statements that define measurable outcomes.

Research diary   A research diary documents the progress of the research process.

Research theme   The features of the data that are deemed relevant to the research phenomenon.

Rigour   Ensuring that research design, methods and conclusions are explicit, public, replicable, open to critique and unbiased.

Risk assessment   The process of identifying hazards that could cause harm and developing strategies to mitigate harm.

Sampling   Selecting a group from a population from whom to collect data from.

Sampling inaccuracies   When the sample does not represent the entire population.

Semi-structured interviews   An interview type that has a predetermined framework of topics that need to be covered but allows for deviation during a research encounter.

Snowball sampling   Involves asking participants currently involved in the research to help recruit new participants into the project.

Structured interviews   A very formalised way of designing and conducting interviews that follows a strict, formulaic script and does not allow for any deviation.

Subjectivity   Research that focusses on the experiences and perceptions of individuals and examines these through the perceptions and interpretations of the researcher.

Summarising   Taking the most important and valuable points from your data and conveying them in your own words.

Synchronous   Events, such as interviews, that are co-ordinated in real time.

Transcription   A written or printed version of an audio recording.

Triangulation   Using multiple methods or data sources to develop a comprehensive understanding of the issue under investigation.

Unstructured interviews   Interviews that they are entirely free-form and open-ended and contain no pre-determined framework or questions.

Video interview   Interviews that are conducted using digital technology, such as video conferencing software.

Voice recorder   The machine or software used to record an interview.

Vulnerability   The state of potentially being exposed to emotional or physical harm.

INDEX

Pages in *italics* refer to figures and pages in **bold** refer to tables.

For Product Safety Concerns and Information please contact our EU
representative GPSR@taylorandfrancis.com
Taylor & Francis Verlag GmbH, Kaufingerstraße 24, 80331 München, Germany

www.ingramcontent.com/pod-product-compliance
Lightning Source LLC
Chambersburg PA
CBHW050649270326
41927CB00012B/2937